Rolling Contact Fatigue in a Vacuum

Michael Danyluk • Anoop Dhingra

Rolling Contact Fatigue in a Vacuum

Test Equipment and Coating Analysis

Michael Danyluk
GE Healthcare
Milwaukee, WI, USA

Anoop Dhingra
Department of Mechanical Engineering
University of Wisconsin
Milwaukee, WI, USA

ISBN 978-3-319-11929-8   ISBN 978-3-319-11930-4 (eBook)
DOI 10.1007/978-3-319-11930-4
Springer Cham Heidelberg New York Dordrecht London

Library of Congress Control Number: 2014953303

© Springer International Publishing Switzerland 2015
This work is subject to copyright. All rights are reserved by the Publisher, whether the whole or part of the material is concerned, specifically the rights of translation, reprinting, reuse of illustrations, recitation, broadcasting, reproduction on microfilms or in any other physical way, and transmission or information storage and retrieval, electronic adaptation, computer software, or by similar or dissimilar methodology now known or hereafter developed. Exempted from this legal reservation are brief excerpts in connection with reviews or scholarly analysis or material supplied specifically for the purpose of being entered and executed on a computer system, for exclusive use by the purchaser of the work. Duplication of this publication or parts thereof is permitted only under the provisions of the Copyright Law of the Publisher's location, in its current version, and permission for use must always be obtained from Springer. Permissions for use may be obtained through RightsLink at the Copyright Clearance Center. Violations are liable to prosecution under the respective Copyright Law.
The use of general descriptive names, registered names, trademarks, service marks, etc. in this publication does not imply, even in the absence of a specific statement, that such names are exempt from the relevant protective laws and regulations and therefore free for general use.
While the advice and information in this book are believed to be true and accurate at the date of publication, neither the authors nor the editors nor the publisher can accept any legal responsibility for any errors or omissions that may be made. The publisher makes no warranty, express or implied, with respect to the material contained herein.

Printed on acid-free paper

Springer is part of Springer Science+Business Media (www.springer.com)

Preface

The motivation to write this monograph was to share engineering experiences related to thin film testing and vacuum systems. Testing of thin film lubricants applied to ball bearings that are used in rotating systems in high-voltage vacuum devices has received little attention in recent decades. This monograph has enabled us to share a unique combination of engineering experiences needed to solve practical problems related to testing of thin film systems. Design and mechanical testing involving vacuum chambers, pumps, plasmas, and the instrumentation needed to operate these has brought to light some less well known combinations of mechanics and physics needed to meet specific testing needs. Indeed, solutions to mechanical engineering problems require creative, cost effective, and innovative combinations of ideas and equipment to achieve sustainable and practical solutions. Necessity is the mother of invention, and this is certainly true when applied to mechanical testing in the context of manufacturing problems.

The impetus behind this monograph is that tribology testing in vacuum in the range of 10^{-5} to 10^{-8} Torr was required to meet a specific testing need. Thin solid film lubrication has been used extensively in rotating anode x-ray tube applications. A non-volatile, carbon-free lubrication, such as silver or nickel-copper-silver, is used to lubricate bearing parts during operation in high vacuum and in the presence of very high voltages, up to 140,000 V. Rotating element bearings that are used in x-ray tubes are a sealed lubrication system, that is, lubrication is added at bearing assembly and the system operates as a closed system for the remainder of its life. The goal to further understand solid lubrication films applied to ball bearings was a financial decision to reduce overall cost and extend bearing life. For example, improving film adhesion will help reduce infantile factory failures. Optimizing film thickness will extend ball-bearing life, and thereby extend the life of the x-ray tube for the customer. However, if the film is too thick it may flake off leading to catastrophic bearing failure. Choice of lubrication is key to prevent interaction with the fragile electronics inside the x-tube. For example, solid lubricants such as silver and gold are stable inside the x-ray tube environment and will not interact with electron beam components during operation. Improvements in film adhesion, deposition techniques, and thickness control occurred in order of priority to add the most value to the customer in the shortest possible time.

The acquired knowledge shared throughout this monograph has been put into practice to correct engineering problems that reduce x-ray tube bearing life, and to deliver the highest value for the customer. The monograph organization has been dictated by necessity to improve ball-bearing system life. The steps to test thin films in rolling contact fatigue under vacuum conditions are outlined in the book. Procedures to prevent contamination and corrupt results are explained to help the reader understand the level of cleanliness required when testing hardware that will be used in a high-voltage and high-vacuum device. Our goal is to discuss and share with the reader issues and problems that may be encountered concerning bearing testing in high vacuum.

The book may be used by students and engineers for guidance to build, test, and commission any vacuum chamber application. Best practices that reduce risk of contamination as well as reduce operating costs are presented in Chap. 2. Diagnostic tools that can detect very slow leaks and that can be used to quantify chamber vacuum integrity are presented in this chapter as well. Within a manufacturing setting, one is often pressured to put a newly constructed vacuum system into use as soon as possible after it is built. However, without proper system commissioning and characterization, the data from the new machine may be corrupt from the very beginning. Time is wasted to either explain anomalies caused by the system or to reproduce all results due to suspect test chamber behavior.

This book is suited for all levels of expertise of scientists and engineers, as well as graduate students in the mechanical and electrical sciences. We start with a general discussion about chamber design, vacuum pumps, and vacuum diagnostic tools. While there are many vacuum-system design books available, in the literature; our goal is to help tailor the vacuum system for testing coatings in rolling contact fatigue (RCF) at high speed and vacuum conditions. We continue in Chaps. 3 and 4 with examples of RCF testing with emphasis on thin film silver lubrication applied to ball bearing steel and silicon-nitride rotating parts. A detailed discussion about the physical vapor deposition ion-plating method is presented in Chap. 6. The ion plating method is well suited for coating a large number of ball bearings at one time with minimal risk of contamination. We use Simulink^{TM} to model and control an ion plating process since is it is readily available to both students and professionals alike.

A survey of the current literature reveals that there is little research concerning process aberration and RCF life. In this book, we use the rolling contact fatigue method to quantify the effects of process aberration on coating life. A discussion about dc-plasmas and their effects on coating performance is presented in Chap. 7. Plasmas are difficult to understand, and there are several text books available concerning the theory and application of plasmas for thin film deposition. We start with practical calculations, such as Debye length and mean-free-path, to help the reader to quantify the plasma related to their process. Our goal is for the reader to understand the plasma as a manufacturing tool using system level inputs such as process pressure and voltage.

In Chaps. 6, 8, and 9, we include a discussion about process control and disturbance rejection. Perhaps more important to a process engineer than a scientist or student, disturbance rejection enables consistent manufacturing coating quality and gives the manufacturing engineer a frame work from which to quantify the effects of process aberration. It is a given that disturbances and even process mistakes will occur in manufacturing coating processes. Choosing the optimal control scheme can help mitigate the effects of unpredictable disturbances on the coating process.

The reader of this book is encouraged to seek out and read further the references cited within the text. Throughout this book, it was our goal to bring together unique combinations of previous engineering work. Rolling contact fatigue testing in vacuum and at high speed is one such example. Upon reading this book, we hope that the reader will be inspired to bring together other such unique combinations of engineering work to help solve problems in engineering and manufacturing.

The authors would like to gratefully acknowledge General Electric Health Care for allowing some of this testing at their facility in Milwaukee, Wisconsin. We would like to thank the University of Wisconsin-Milwaukee for use of its computing and laboratory resources between Fall 2010 and Spring 2014.

Milwaukee, WI, USA
Michael Danyluk
Anoop Dhingra

Contents

1 Introduction 1
   1.1 Coating Processes Compatible with High-Voltage Devices 2
   1.2 Monograph Overview 3
   1.3 Plasma Diagnostics and Measurements 4
   1.4 Process Control Considerations 4
   1.5 Monograph Organization 4
   References 5

Part I Vacuum Systems Infrastructure and Chamber Design 7

2 Vacuum Chamber Design 9
   2.1 Introduction 10
   2.2 Vacuum Chamber Design Considerations 10
   2.3 Chamber Material Selection 13
      2.3.1 Material Outgassing 13
      2.3.2 Vacuum Welds, Gaskets, and Attachments 14
      2.3.3 Vacuum Chamber Isolation 14
   2.4 Vacuum Components Selection 15
      2.4.1 High-Vacuum Pumping System 15
      2.4.2 Vacuum Measurement System Selection 15
      2.4.3 Vacuum System Safety Interlocks 16
   2.5 Case Study: Chamber Vibration Isolation and Measurement 17
      2.5.1 Vibration Transmissibility Design 17
      2.5.2 Accelerometer Response Comparison 21
   2.6 Case Study Continued: Optimum Chamber Design 24
      2.6.1 Design Optimization Analysis 25
      2.6.2 Compare Optimum Designs 30
   2.7 Assembly and Cleaning 31

	2.8	Residual Gas Analysis	32
		2.8.1 Establishing Chamber Baseline	32
		2.8.2 Helium Leak Check	33
	2.9	System Commissioning	33
		References	34
3	**Rolling Contact Testing of Ball Bearing Elements**		35
	3.1	Introduction	36
	3.2	Analysis of Rolling Contact in Vacuum	39
		3.2.1 Calculations of Contact Stress	39
		3.2.2 Contact Load Versus Cycles	46
	3.3	Rolling Contact Wear	46
		3.3.1 Empirical Approach	47
		3.3.2 Thermodynamic Approach to Friction and Wear	49
		References	51
4	**Rolling Contact Fatigue in High Vacuum**		53
	4.1	Introduction	54
	4.2	Rolling Contact Fatigue Test Platform	55
		4.2.1 Test Configuration in Oil	59
		4.2.2 Test Configuration in Vacuum	60
		4.2.3 Rolling Elements in Vacuum	61
	4.3	Rolling Contact Fatigue Vacuum Test	64
		4.3.1 Coated Rolling Elements in Vacuum	64
		4.3.2 Test Assembly in Vacuum Chamber	65
		4.3.3 RCF Test Failure Criterion	65
		4.3.4 Analysis Tools	66
		4.3.5 RCF Test Results	70
		4.3.6 Post-Test Autopsy of Contacting Elements	72
		4.3.7 Post-Test Elemental Content of Film	76
	4.4	Friction and Wear Calculations	78
		4.4.1 Third-Body Transfer Mechanism	78
		4.4.2 Empirical Comparison: Lundberg–Palmgren Model	82
		References	85
5	**Coating Thickness Calculation and Adhesion**		87
	5.1	Thickness Measurement Techniques	88
		5.1.1 Calculate Thickness by Weight	89
		5.1.2 Thickness Measurement Using XRF Spectrometry	91
	5.2	Pretest Adhesion Check	92
		5.2.1 Scratch Test Ball Sample	92
		5.2.2 Particulate Detection Tape Testing	93
	5.3	Closing Comments	95
		References	95

Contents

Part II Simulation and Testing of Thin Films in a Vacuum Environment 97

6 Ion-Plating Process Model 99
   6.1 Plasma and Deposition Processes 101
   6.2 Postdeposition Fatigue Testing 103
   6.3 Plasma-Assisted Deposition 105
   6.4 Coating Procedure for RCF Testings 107
   6.5 Plasma Effects on Coating Thickness 109
   6.6 Analysis of Extreme DoE Coating Tests 110
   6.7 RCF Testing of Extreme DoE Coated Balls 112
   6.8 Ion-Plating Model in Simulink™ 114
      6.8.1 Cathode dc-Sheath Model 115
      6.8.2 Sputter and Evaporation-Sputter Deposition Models 116
      6.8.3 Elements of the Ion-Plating Simulation Model 118
   6.9 Process Model Simulation 120
   References 125

7 Effects of Process Parameters on Film RCF Life 127
   7.1 Plasma Diagnostic Tool 128
      7.1.1 Langmuir Probe Experimental Procedure 129
      7.1.2 Choosing an Appropriate Diagnostic Model 131
      7.1.3 Correct Interpretation of Langmuir Probe Data 132
   7.2 Plasma Effects on Film Properties and Composition 135
      7.2.1 Ion and Elemental Mixing 135
      7.2.2 Auger Electron Spectroscopy Test 135
      7.2.3 Ion and Element Implantation Modeling 136
   7.3 Film Properties Calculation 137
      7.3.1 Layered Film Structure Properties 138
      7.3.2 Film-Stress Calculation 139
   References 144

Part III Control and Disturbance-Rejection of Thin Film Deposition Systems 145

8 Real-Time Process Control 147
   8.1 Experimental Setup Using Simulink Real Time 147
   8.2 Plant Model Characterization 149
      8.2.1 Conductance Valve Characterization 151
      8.2.2 Manometer and Plasma-Current Response 153
   8.3 Process Control Using Plasma Current 155
   Reference 155

9	**Closing Chapter: Disturbance Rejection**	157
	9.1 System Identification	158
	9.2 Model Predictive Control	162
	9.3 Closing Comments and Future Testing	164
	References	164
Index		165

Chapter 1
Introduction

Testing of solid thin film lubricants for tribology applications that operate in vacuum presents unique challenges not encountered when testing in air. The end use of the lubricant film will determine the level of vacuum needed for testing. For example, if the application use of the lubricant requires high-vacuum conditions, then the film should be tested in those conditions as well. Testing in the milli-Torr range is significantly easier than testing in high-vacuum conditions, so long as the presence of some gas is permissible for accurate evaluation of the film. If the system under test will not have continuous pumping and still requires high-vacuum operating conditions, then all gases from the coating and hardware components inside the test chamber must be removed before testing begins. For example, if the system and coating under test will be sealed, that is, isolated from the vacuum pumping system, then all outgassing must take place before isolation from the pumping system. Rotating anode x-ray tubes, which are high-voltage and high-vacuum devices and require a high-vacuum operating environment, do not operate with constant pumping, and it is therefore required to have all components sufficiently outgassed before disconnecting from the pumping system. However, outgassing all components prior to testing requires significant preparation. Alternatively, all components required for testing coatings would not need to be fully outgassed provided that continuous vacuum pumping is used to remove outgassing molecules during the test.

Thin films deposited with argon plasma assist require milli-Torr (10^{-3} Torr) pressure range processes and may contain trapped gases when operated in high vacuum conditions. The typical pressure range for high vacuum processes is 10^{-5}–10^{-8} Torr, which is three to five decades lower in pressure magnitude than what the film deposition process requires. For example, the operating pressure of silver or lead solid lubricant used on an x-ray tube ball bearing is 10^{-7} Torr, yet these films are deposited in the milli-Torr range. Trapped gas molecules such as argon and nitrogen will escape from the coating structure when the operating pressure is lower than the deposition pressure. The x-ray tube environment is unique in that the vacuum space must be free of gaseous contaminates to prevent high-voltage arcing. Even the presence of inert gas molecules is detrimental to the operation of the device. The challenge for production engineers is to use industry-standard coating practices to reduce cost, which requires film deposition applied in the milli-Torr range, and still ensure a gas-free environment for x-ray generation. For this application, the coating itself must be outgassed prior to operation of the x-ray tube.

This monograph shares our engineering experiences related to thin film testing and vacuum systems used in a manufacturing setting and is divided into three broad areas: (i) vacuum systems with emphasis on chamber design and infrastructure, (ii) simulation and testing of thin films in a vacuum environment with accompanying mathematical modeling, and (iii) introductory control and disturbance rejection in thin film deposition systems. This collection of topics came out of necessity to improve the rolling contact fatigue (RCF) life of thin solid film lubricants applied to ball bearings operated in high vacuum. Why do some coatings last longer than others coming from the same ion-plating machine and process? Observation that pressure and voltage aberration, along with gas contamination that occurred during application of the coatings to the ball surface lead us to investigate how coating life was being affected. How would one quantify the impact of process aberration on coating performance in a systematic and repeatable way? These topics and more will be covered in the following chapters of this book.

1.1 Coating Processes Compatible with High-Voltage Devices

Lubrication coatings for bearing systems that operate at high rotational speeds and high temperature in high vacuum conditions rely heavily on thin solid film lubricants. If graphite or other carbon-based lubricants cannot be used, then thin film silver or lead may be used as a lubricant if the operating temperature of the coating is between 200°C and 500°C. Ball bearing systems that use thin solid film lubrication and operate in voltage range up to 120 kV must use non-particulate generating films to prevent arcing and high-voltage breakdown during operation. Historically, thin films of silver, copper, lead, gold, and molybdenum disulfide have been used in rotating anode x-ray tubes. These films are inert with respect to the operating conditions inside the tube and with tolerable amounts of particulate generation.

Physical vapor deposition (PVD) and magnetron sputtering are two processes commonly used to deposit thin films on a substrate. PVD systems are a collection of subsystems that run concurrently within the process. For example, the plasma voltage subsystem, deposition sources, the pressure monitor, all may operate independent of each other at the subsystem level. Any lack of communication between these subsystems makes the coating process more sensitive to process aberration. For example, the plasma associated with a PVD process can be interrupted by either contamination or high-voltage isolation collapse within the process chamber. Both of these interruptions affect deposition microstructure and film stress and ultimately the rolling contact fatigue life of the film.

The PVD ion-plating process is the focus of this book. Ion plating is a momentum transfer process in which the kinetic energy of the process gas ions is transferred to the deposition material through atomic collision. In this process, argon ions collide with neutral atoms that are to be deposited, such as silver and copper that have entered the plasma sheath from either an evaporation or sputter source. The energy transferred from the argon ions to the silver atoms, for example, is sufficient to implant the silver atoms into the lattice structure of the ball material. The energy transfer mechanism of this collision, along with its effects on coating RCF life, will be discussed in Chaps. 6 and 7.

1.2 Monograph Overview

This monograph is intended as a guide for construction and testing of RCF systems for high vacuum operation as well as process development and control of thin film deposition systems. The ion-plating process and the dc plasma associated with it were chosen for study in this text since these are mature processes and have been explained by industry and academic experts such as Mattox (1998) and Lieberman and Lichtenberg (1994) to name a few. Based on the authors' experience, the ion-plating process is the cleanest concerning silver coatings on ball bearings that will be used in high-voltage and high-vacuum devices such as rotating anode x-ray tubes.

When developing a coating process for ball bearings used in high-voltage and high-vacuum devices, there is a trade-off between (i) film adhesion, (ii) film purity, (iii) and film composition for multielement films. For example, ion plating a nickel-copper-silver film using 2 keV argon ions promotes good adhesion and neutral-atom implantation of a nickel-copper-silver to the ball surface. However, the ion flux associated with the voltage and pressure that is required to accelerate the argon ions to 2 keV energy can at the same time cause buckling and yielding within the film. Ion energy greater than 500 eV can induce local thermal melting within the film which helps to relax internal film stress, Mayr and Averback (2003). However, the argon ions at this energy cause interlayer mixing which results in film interlayer contamination as well, Waits (1998). These two situations, local thermal melting and interlayer mixing will be explained further in the book. The RCF method in vacuum will be used to quantify the effects of each phenomena.

Current RCF test methods require testing with oil-based lubrication. However, testing with oil lubrication does not represent the operating environment of the thin film system inside a high-voltage and high-vacuum device such as an x-ray tube. Scratch testing and hardness test methods do not test the film in high-cycle fatigue loading, which renders them useless concerning rotating anode x-ray tube applications. A unique test platform that can test coating adhesion and life under RCF conditions in vacuum and at high rotational speeds is developed and presented in Chap. 4.

1.3 Plasma Diagnostics and Measurements

Understanding process conditions during application of thin films requires unique measurement tools. Concerning PVD processes, it is essential to know the plasma properties as a function of process voltage and pressure in order to control film structure. A Langmuir probe may be used to measure and calculate plasma properties in the quasi-neutral region of the plasma during the deposition process. With these properties, one can calculate ion kinetic energy, plasma sheath thickness, and electron current very near the surface of the ball as described in Lieberman and Lichtenberg (1994) and in Ruzic (1994). Knowledge of plasma properties is essential to understand the implantation depth of deposited materials such as nickel, copper, and silver.

1.4 Process Control Considerations

Processes requiring plasma interaction with substrate materials prior to deposition pose unique control challenges. The difficulties lay not in choosing an effective control algorithm for a given deposition process. Rather, challenges related to sensor locations and observability of plasma state variables for feedback control pose the greatest hurdle for improving control of deposition processes. For example, typical pressure monitoring systems used inside vacuum chambers have slow response times, on the order of 1–2 s. Yet, a pressure disturbance within the plasma, due to contamination, for example, takes place in less than 2 s. A more sophisticated solution to reduce coating process variation involves improved sensor input to feedback control schemes. Using information from in situ sensors and monitors, a control algorithm may be tuned to mitigate drift and disturbance inputs independent of process history.

1.5 Monograph Organization

The focus of this book is rolling contact fatigue testing under high vacuum conditions. The RCF in vacuum is used to establish a connection between ion-plating process parameters and coating RCF life.

The monograph is divided into three parts. The first part, Chaps. 2, 3, 4, and 5, deal with vacuum systems and in particular the design of vacuum chambers and associated infrastructure. Chapter 2 covers aspects of vacuum chamber design related to vibration isolation, vacuum pumping system selection, and system characterization using residual gas analysis (RGA). Vibration detection at coating failure is required for all RCF testing in vacuum. Therefore, rigorous vibration testing and numerical simulation design procedures are presented to quantify vibration transmittance through the chamber structure.

An introduction to the necessary calculations for rolling contact and wear contact is presented in Chap. 3. A complete presentation of engineering tribology and wear may be found in texts such as "Introduction to Tribology," by Bhushan (2002), and "Engineering Tribology," by Stachowiak and Batchelor (2005). Chapter 4 introduces a RCF test rig for high rotational speed testing under high-vacuum conditions. The RCF platform in high vacuum is validated for two ball sizes and coating systems and materials. A third-body storage model is applied to experimental data with good agreement between modeling and test results. Low-cost procedures to estimate average coating thickness and to quantify adhesion of coated balls are presented in Chap. 5.

The second part of the monograph, Chaps. 6 and 7, deals with simulation and testing of thin films in a vacuum environment. Chapter 6 presents a numerical model of an ion-plating process. The process is simulated using Simulink^{TM} with subsystem models taken from the thin film literature related to (i) Hertz-Knudsen evaporation, (ii) a matrix-sheath sputtering and implantation, and (iii) a linear systems pressure control models. Chapter 7 discusses the trade-offs associated with process pressure and voltage sensitivity. Specifically, the pressure and voltage effects on coating composition and deposition physics are studied using Auger electron spectroscopy (AES) and numerical modeling. A diffusion-based, stress and relaxation defect model is used to calculate film stress as a function of pressure during deposition. There is good agreement between the stress and implantation models with RCF life test results over the pressure and voltage ranges tested.

In the third part, Chap. 8, applies hardware in the loop testing to the ion-plating system of earlier chapters. Evaluation of controller feedback information from two sources, (i) a pressure manometer attached to the chamber and (ii) plasma total current measurements through the dc power supply and a Langmuir probe, is discussed. Finally, Chap. 9 presents some concluding remarks related to disturbance rejection using model predictive control.

References

Bhushan B. Introduction to tribology. New York: Wiley; 2002.
Lieberman M, Lichtenberg A. Principles of plasma discharges and materials processing. 2nd ed. New York: Wiley; 1994.
Mattox D. Handbook of physical vapor deposition processing. Westwood: Noyes; 1998.
Mayr SG, Averback RS. Effect of ion bombardment on stress in thin metal films. Phys Rev B. 2003;68(214105):1–10.
Ruzic D. Electric probes for low temperature plasmas. New York: AVS Press; 1994.
Stanchowiak G, Batchelor A. Engineering tribology. Burlington: Elsevier Butterworth-Heinemann; 2005.
Waits RK. Thin film deposition and patterning. New York: AVS Press; 1998.

Part I
Vacuum Systems Infrastructure and Chamber Design

Chapter 2
Vacuum Chamber Design

Abbreviations

AFM	Atomic force microscope
EB	Electron beam
Hz	Hertz cycles per second
K	Thermal conductivity
L/s	Liters per second
NO/NC	Normally open/normally closed
PVD	Physical vapor deposition
RCF	Rolling contact fatigue
RF	Radio frequency
RGA	Residual gas analysis
RPM	Revolutions per minute
Sccm	Standard cubic centimeters per minute
VAC	Volts alternating current

A discussion of mechanical testing in high vacuum is presented in this chapter. Vacuum component selection and test chamber materials are discussed with emphasis on what is needed for testing high speed rotating systems in vacuum similar to a rotating anode inside an x-ray tube device. Low speed testing of rotating systems to detect balance shift is also discussed with emphasis on vibration transmittance through key vacuum components. Two case studies are explored that highlight the balance between cost and complexity related to chamber design and its long-term operation.

2.1 Introduction

The first step in designing a vacuum chamber is to determine how the chamber will be used and to have a clear understanding of the processes that will be carried out inside the chamber. The designer must be aware of the risks for contamination to the vacuum chamber, and how will the inside of the chamber be checked for cleanliness prior to starting a process. Some additional design consideration includes what types of coatings are present inside the chamber during the test, at what temperature, and at what pressure will the test be carried out. Will the process involve high voltage, that is, application of high-voltage bias, during the process? Without a clear understanding of the intended use and fine details of the test, you may inadvertently purchase equipment that is improperly sized for your needs. Worse yet, you may purchase incorrect components which will require costly rework of the equipment.

Maintenance costs, along with system performance, need to be considered when selecting vacuum chamber components. For example, a cryogenic vacuum system is very clean and requires little hands-on maintenance. However, it does require uninterrupted facility power to run the helium compressor, usually 480 VAC 3 phase, for example, even when the chamber is vented to the atmosphere. The cold head of the cryogenic system is not at all tolerant of hydrocarbon contamination. Oil and even low vapor pressure vacuum grease can destroy the cold head of a cryogenic pumping system. In comparison, a turbo pumping system can provide the same pumping speed but with a smaller foot print than a cryogenic system. Turbo pumping systems are less affected by hydrocarbon contamination, but are at risk in the event of a rapid venting process. Turbo pumps may be severely damaged by rapid pressure changes within the chamber and may fail catastrophically if the chamber gas load is too high. Each pumping system has its advantages and disadvantages, and the designer needs to decide what level of risk is tolerable concerning their process and who will operate the equipment.

A proper organization and arrangement of components inside the chamber can reduce contamination risk and shorten maintenance time. For example, tight corners and layered structures inside the chamber should be avoided where possible. Layered structure such as shelves or closely spaced plates allows contamination and debris to collect without easy visual detection. All fixtures and components should be easily removable, requiring simple hand tools and little effort and skill for cleaning.

2.2 Vacuum Chamber Design Considerations

The pumps and reciprocating hardware associated with vacuum systems can generate significant mechanical vibrations that are often transmitted to the chamber. These systems require hermetic and structural connection between components

2.2 Vacuum Chamber Design Considerations

Fig. 2.1 Types of bellows: (**a**) welded, (**b**) hydroformed

which can increase vibration transmittance. However, if one separates the hermetic and structural requirements, a design solution may be achieved. For example, the welded-flexible bellows shown in Fig. 2.1a enables a hermetic seal with very little vibration transmittance, but does not support structural load. In comparison, the formed bellows of Fig. 2.1b is less flexible than the welded design and can support light to moderate loads and is not as rigid as a bolted gasket joint. The welded bellows may be used, for example, between the pump and chamber to minimize vibration transmittance from the pump. The pump may then be supported externally from the chamber to prevent vibration transmittance. For example, one novel idea from Iijima (1982) allows atmospheric pressure to offset the weight of the pump and removes fixed bracket attachments to the chamber, thereby minimizing vibration transmittance from the pump.

The stiffness of a bellows assembly may vary depending on how it was manufactured. The two most common manufacturing methods are hermetic welding and hydroforming, both shown in Fig. 2.1. The hydroformed bellows are much stiffer than the welded design and have some load carrying capability. In comparison, the hermetic-weld bellows offer the best vibration isolation but are not capable of supporting a load of any kind. In fact, the welded bellows requires external support fixtures to be effective. The welded bellow consists of a series of thin flat disks, usually at least 20, that are welded together at the inner and outer diameters. The flexibility of each disk provides the vibration isolation.

Chamber material and geometry should be chosen to minimize chamber outgassing time and maintenance. The surface finish of the chamber walls may influence the test inside the chamber. For example, engineers designing a chamber to carryout particle physics experiments similar to Baglin et al. (1998) would be concerned with how the chamber wall surfaces will interact with the electron cloud during their test. If electron-beam or radio-frequency heating is going to be used to heat test components, the surface finish and material of the chamber walls may

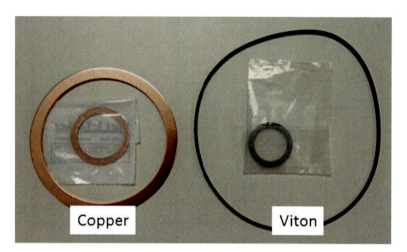

Fig. 2.2 Copper and Viton rubber gaskets

interact with those heating sources, which could result in unintended heating of the test sample inside the chamber or, worse, direct heating of the chamber itself.

Temperature isolation may also influence chamber material selection. For engineers designing a vacuum chamber furnace for a fluxless soldering process as in Lee et al. (2006), minimizing heating time and operation costs was most important. When applying heat inside a vacuum chamber, the primary heat dissipation path will be conduction through the support fixtures and connections to the chamber structure. A simple conduction calculation assuming heat flow through a flat plate $H = kA\Delta T/L$ can give insight about how material choice and geometry will affect the test inside the chamber. For example, the thermal conductivity, (k), for metals is much larger than that for ceramics. If the process involves significant heating of 500 °C or more, then ceramic fixtures or even ceramic liners may be needed to protect the chamber against long-term damage. If heat is generated within the test hardware itself, in case of rotating internal friction, then chamber external connections should be isolated to allow use of less expensive and flexible Viton[TM] o-ring seals instead of rigid copper gaskets. A comparison of Viton and copper gaskets is shown in Fig. 2.2. One reason for using metal gaskets, copper, or silver-coated copper is to minimize outgassing time when the chamber is under high vacuum. Copper gaskets will outgas more quickly than Viton, and Viton gaskets will outgas much quicker than silicon gaskets, for example. Why is outgassing important? It is important because the gaseous molecules and atoms leaving the gasket material will interact with test sample while under vacuum.

Thermal isolation reduces total energy consumption for a test and prevents damage to heat-sensitive components that may be mounted on the chamber exterior. The researchers in Lee et al. (2006) were able to drive their sample temperature to 400 °C while maintaining chamber base exterior temperature of only 55 °C.

2.3 Chamber Material Selection

Another consideration in chamber design is determining what level of vacuum is needed for the test or process. It is significantly more work and a lot more expense to reach high-vacuum condition $<10^{-5}$ Torr than it is to test in the rough-vacuum 10^{-4}–10^{-2} Torr range. Vacuum processes that occur in the 10^{-4}–10^{-2} Torr range are commonly referred to as "milli-Torr" range vacuum. For example, if the process involves any type of voltage biasing or high-voltage application, then at least high-vacuum conditions will be required. At pressures in the milli-Torr range, there are sufficient number of gas atoms and molecules present inside the chamber to be ionized when high voltage is applied. The ionization is a result of leakage current through the gas during the test. Such leakage current paths can alter your test results and may cause significant damage to the equipment. On the other hand, if the process or test requires only an inert and evacuated chamber environment, that is, no high voltage and no heat, then rough vacuum in the 10^{-2}–10^{-4} Torr range may be sufficient.

The focus of this chapter is vacuum chamber design for testing components used in high-voltage and high-vacuum devices. One such example is a rotating anode x-ray tube device. The chamber environment must be sufficiently outgassed to prevent ionization and leakage current during operation. For example, the chamber could be in the 10^{-7} Torr range, but may not be sufficiently outgassed for 120 kV operation. The chambers used for component testing should have the same operational characteristics as the end-use application.

2.3 Chamber Material Selection

The most common and recommended material for high-vacuum chamber applications is 304 stainless steel. Quick surface outgassing and resistance to surface oxidation are among the most beneficial aspects of using 304 stainless steel (304SS). However, this material can be difficult to machine and may require frequent tool and cutting bit replacement. Stainless steel chambers also transmit vibration to all attached components. Vibration from pumping systems and those transmitted from the floor can corrupt sensitive vacuum instruments as well as confound vibration measurements of the hardware under test inside the chamber.

2.3.1 Material Outgassing

Material surface outgassing characteristics should be considered when choosing vacuum chamber material. For example, 6061 aluminum is hermetic and light weight and will provide necessary structural support for a vacuum chamber application. However, due to its chemical composition, an aluminum surface will collect large amounts of surface moisture each time the chamber is opened to the atmosphere. If the process inside the chamber is sensitive to moisture, nitrogen, and oxygen, the process should not be started until the chamber is sufficiently outgassed. The surface moisture will need to be removed each time the chamber is opened, resulting in long pump-down times and delays in starting the process. In contrast, 304SS collects very little moisture, thus allowing faster outgassing and improved process cycle time in a production environment. The outgassing characteristics for candidate chamber materials are tabulated in Mattox (1998).

2.3.2 Vacuum Welds, Gaskets, and Attachments

All welded joints that seal the chamber from atmosphere need to pass a leak rate test to $10^{-9} \frac{\text{mBar} \cdot \text{Liter}}{\text{Sec}}$ and should be constructed in such a way as to prevent trapped gas or accumulated debris. The weld must provide sufficient strength to support measurement instruments and also have flexibility to prevent fatigue damage after repeated evacuation and venting. Atmospheric pressure on the chamber during vacuum is 14.6 psi at sea level; therefore, large flat sections within the chamber should be avoided due to repeated bending and flexure at welded joints each time the chamber is pumped down. Corners and cavities must be properly vented or eliminated whenever possible. For example, a blind hole inside the chamber can trap gas if it is not properly vented. With the threaded fastener sealing off the opening of the hole, trapped gas will bleed into the chamber, resulting in a virtual leak and long pump-down time. Vented fasteners should be used on all bolted joint assemblies within the vacuum space of the chamber. Press-fit assembles inside the chamber can cause virtual leaks as well if not properly vented. In general, make sure all components, fixtures, fasteners, and gaskets are properly vented before installing them on the chamber.

2.3.3 Vacuum Chamber Isolation

Vibration isolation from components outside the chamber may be achieved using a flexible bellows similar to the one shown in Fig. 2.1. For example, if vibration measurements are needed while the test hardware is under vacuum, the chamber will need to be isolated from vibration input from vacuum system components. The hardware required to maintain vacuum can be stiff and heavy, resulting in significant vibration transmittance to the test element(s) inside the chamber. If chamber components heat up during the test, then the vibrational characteristics of the chamber will change and could confound test measurement results. For example, if one wishes to track the peak vibration amplitude of a rotating system as it is heated under vacuum, then any vibration artifacts from the chamber must be accounted for in order to have confidence in the measurement. If temperature isolation and control is required, the chamber system may require active cooling by forced convection or possibly temperature-controlled process fluid through the chamber structure.

2.4 Vacuum Components Selection

This section is intended as an overview of key components that are needed to support a high-vacuum test chamber. More detail on these components as well as many others may be found in vacuum technology textbooks such as by Mattox (1998) and O'Hanlon (1989). When selecting vacuum components, it is best to work with a brand-name equipment supplier who can help you to make the best choice.

2.4.1 High-Vacuum Pumping System

Cryogenic vacuum-pumping systems are among the cleanest in the industry for use for high-vacuum chambers. A cryogenic vacuum pump is hermetically sealed from outside atmosphere once the vacuum chamber has been evacuated to less than approximately 10^{-3} Torr. Another advantage of the cryogenic system is that there are no moving parts inside the vacuum space so that in the event of a catastrophic leak, or an accidental rapid venting of the chamber, the pumping system will not be damaged. In contrast, a turbo pump is a momentum transfer pumping system. An array of compressor blades attached to a rotor drum and bearing assembly is rotated to about 75 kRPM to force or evacuate gases out of the chamber. Both turbo and cryogenic pumping systems are capable of reaching high vacuum, 10^{-7} Torr. In general, turbo pumping systems are more economical and require less space than cryogenic systems.

2.4.2 Vacuum Measurement System Selection

Vacuum measurement systems are selected based on process sensitivity to pressure and on the method used to control chamber pressure. For example, a thin film PVD system may require chamber pressure control during deposition. Usually two gauge systems are required: one for high-vacuum 10^{-5}–10^{-7} Torr and one for rough-vacuum 10^{-3}–10^{-4} Torr. The gauge controllers from each system provide dry contact relay circuits to enable safety interlocks and process protection. The relay circuits are either normally open (NO) or normally closed (NC). These circuits are different from digital communications in that these relays interface well with a large variety of industrial equipment and do not require specialty connector hardware such as RJ45 or RS-232, 485. Processes in the milli-Torr range, 10^{-3}–10^{-2} Torr, require a self-heating manometer pressure sensor to give the most accurate pressure measurement during the process. For processes less sensitive to pressure, and occurring in the ultrahigh-vacuum regime such as 10^{-5} Torr or less, either a hot filament or cold cathode magnetron gauge would be sufficient.

2.4.3 Vacuum System Safety Interlocks

All vacuum subsystem components have internal safety interlock relays. Safety interlocks apply to three aspects of the chamber system operation: (i) safety of the operator, (ii) protection of the equipment, and (iii) protection of the test or part inside the chamber. Operator safety is the most important aspect of system interlock configuration. Vacuum system processes often bring together voltage and current in the ranges of 10–2,000 V and current as high as 15 amps. Turbo pumps rotating at 75 kRPM, as well as cryogenic pumps that cool to 17 K, may be very dangerous to an operator if they malfunction. Consideration of the rotating or friction-generating devices inside the test must be included as well when designing your chamber. What if the device is rotating very fast and fails, what is the containment capability of your chamber? What if the operator tries to open the chamber before the process is complete? Use of safety interlocks will enable safe operation of the chamber system in the event of operator error or impatience.

Protection of system equipment and the test parts inside the chamber is also very important. What if there is a power or other utility outage during the process? The objective is to bring the system to zero energy as fast as possible without damaging test components inside the chamber. If the chamber is vented during a high-heat situation of the process, it could damage the parts and components inside, rendering some instruments a complete loss if not properly interlocked with a control system. Figure 2.3 contains a safety interlock circuit that protects both operator and test rig. For example, if the RCF chamber is vented while the test is still rotating, the pressure interlock will open, which will shut down all systems immediately. If the system is running unattended and there the process water cooling flow stops, the system will shut down to prevent overheating. Referring to Fig. 2.3, the high-voltage bias heating to the RCF test rod is enabled only with the pressure threshold is below milli-Torr range, to protect the operator from injury while the chamber is open.

Presented next is a case study dealing with design and optimization of a vacuum chamber with heated and rotating components. Three chambers were designed, tested, and optimized for this application. Optimizing an existing chamber design enables one to focus on design aspects that will give the most appropriate solution for the problem under consideration. For example, if the structure and geometry of a tool or piece of equipment is considered expensive to build and to operate, one can use design optimization techniques to construct an objective function to track the cost of the tool or equipment over its projected operating life. In contrast, optimization on a design concept before it has been built is more challenging because all

2.5 Case Study: Chamber Vibration Isolation and Measurement   17

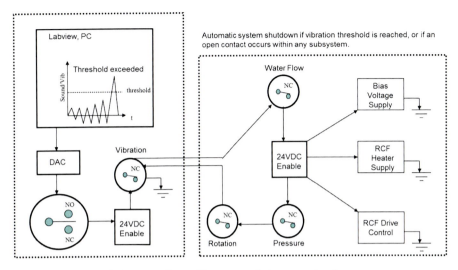

Fig. 2.3 Safety interlock system for a rolling contact fatigue test chamber

of the possible trade-offs are not yet established. The ideal situation is to apply optimization techniques to a prototype design and verify conclusions and assumptions with sub-element testing; then return to update and complete the optimization task using the lessons learned from the first tier-optimized prototype design.

2.5 Case Study: Chamber Vibration Isolation and Measurement

This section and the next present three vacuum chamber test rigs that were designed to measure vibration of rotating components in high vacuum and at high temperature. The designs were optimized using MATLAB Optimization Toolbox, for vibration isolation, cost of operation, and protection of the sensitive measurement tools attached to the chamber. Where ever possible, the systems were constructed using off-the-shelf components from vacuum hardware suppliers. As a requirement, the chambers must be assembled by engineers and technicians using common hand tools with basic mechanical laboratory skills. No specialized training is required, a key factor in keeping maintenance costs low.

2.5.1 Vibration Transmissibility Design

Testing in vacuum presents unique challenges not encountered when testing in air. Vibration isolation in vacuum can be particularly challenging due to the stiffness

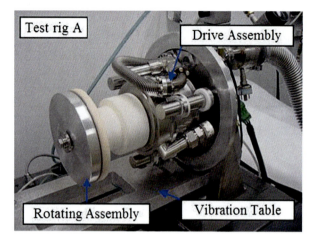

Fig. 2.4 Test rig A: configuration of the drive, rotating, and table assemblies in air (Reproduced with permission from Rev. Sci. Instrum. 82, 105113 Copyright 2011, AIP Publishing LLC)

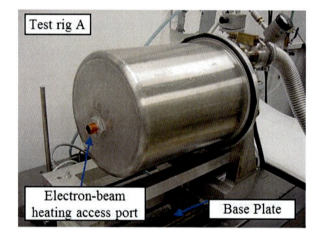

Fig. 2.5 Test rig A under vacuum with electron-beam heater port shown (Reproduced with permission from Rev. Sci. Instrum. 82, 105113 Copyright 2011, AIP Publishing LLC)

properties of materials such as 304 stainless steel. Three vacuum chamber systems are compared in this section. The first test rig, test rig A, is presented in Fig. 2.4 and includes a rotating aluminum–ceramic braze assembly. For this test, the brazed assembly is rotated at 15 Hz in vacuum while being heated to 200 °C.

Vibration data as a function of time and temperature is collected and used to assess the structural integrity of the brazed assembly. Heat is applied using a coiled tungsten filament at 5.5 amps and biased to 1 kV using a dc power supply. The biased coiled filament is an electron-beam heating system, with connection to the power supply through the copper tube as shown in Fig. 2.5.

Test rig A was also used to characterize a welded-flexible bellows assembly for application in the RCF high-vacuum test rig. The goal of this test was to measure vibration transmissibility through the bellows assembly and to compare it with a fixed-flange design. From Fig. 2.4, it can be seen that the drive and vacuum vessel assemblies are mounted independent of the rotating assembly and vibration table.

2.5 Case Study: Chamber Vibration Isolation and Measurement

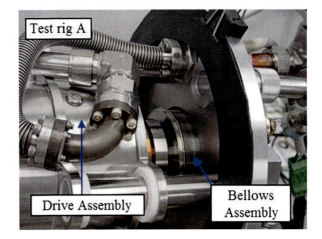

Fig. 2.6 Close-up of test rig A to point out the weld-flexible bellows assembly (Reproduced with permission from Rev. Sci. Instrum. 82, 105113 Copyright 2011, AIP Publishing LLC)

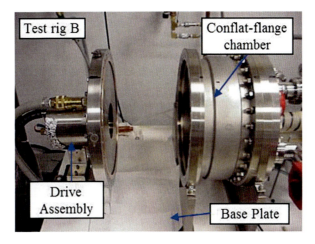

Fig. 2.7 Test rig B uses an off-the-shelf conflat flange joint vacuum chamber (Reproduced with permission from Rev. Sci. Instrum. 82, 105113 Copyright 2011, AIP Publishing LLC)

Specifically, the bearing on the rotating assembly is rigidly attached to the vibration table. The table is supported on two displacement sensing probes. The probes record displacement during rotation of the aluminum–ceramic component spinning in vacuum. The welded-flexible bellows is used to seal the vacuum system between vibration table and drive assembly. The bellows assembly is presented in Fig. 2.6. One end of the bellows is attached to the drive assembly, while the other end is connected to the mounting interface on the bearing of the rotating aluminum–ceramic part.

Test rig B is shown in Fig. 2.7 and does not include the welded-flexible bellows assembly. The rotating assembly is rigidly attached to the drive assembly, and therefore the vibration measurement is confounded with external inputs from the vacuum pumps. Accelerometers are used to measure the vibration transmission from the drive assembly of rigs A and B to the measurement table.

Fig. 2.8 Accelerometer locations on test rig A (Reproduced with permission from Rev. Sci. Instrum. 82, 105113 Copyright 2011, AIP Publishing LLC)

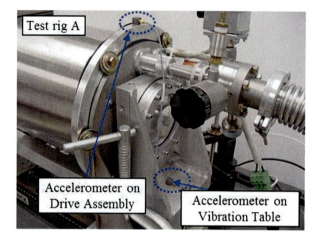

Fig. 2.9 Accelerometer locations on test rig B (Reproduced with permission from Rev. Sci. Instrum. 82, 105113 Copyright 2011, AIP Publishing LLC)

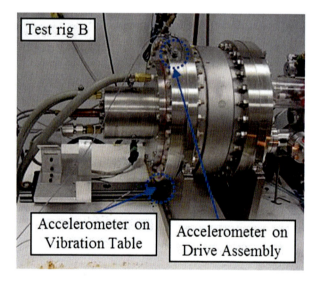

Figures 2.8 and 2.9 illustrate the accelerometer measurement locations for each test rig A and B. The location of each accelerometer on the test rigs A and B is the same with respect to the drive assembly and vibration table. The point of comparison between the two test rigs is the vibration transmittance between accelerometers for each rig.

The vibration transmittance may be modeled as a collection of springs and masses, one for each component of the test rigs. For example, Fig. 2.10 illustrates the connectedness of the rotating assembly, the drive assembly, and the vibration measurement table. An excitation force $F_2(t)$ originating from an imbalance of the rotating system is applied to the vibration table. The subsequent displacements of the vibration table and drive assembly, $x_2(t)$ and $x_3(t)$, are measured using the

2.5 Case Study: Chamber Vibration Isolation and Measurement

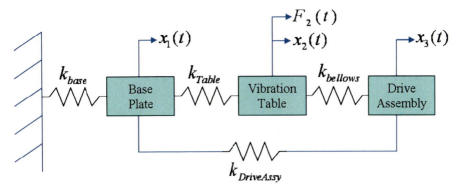

Fig. 2.10 A spring–mass system model illustrating the force transmissibility test for test rig A (Reproduced with permission from Rev. Sci. Instrum. 82, 105113 Copyright 2011, AIP Publishing LLC)

accelerometers presented in Fig. 2.8 and compared with the vibration response of Fig. 2.9.

For the spring–mass model shown in Fig. 2.10, two test conditions for vibration transmittance are considered: (i) in high vacuum and (ii) in air with the bellows detached from the vibration table. The stiffnesses, k_{Table} and $k_{DriveAssy}$, are nonzero for both test conditions since both the vibration table and drive assembly are present for each test condition (i) and (ii). The bellows stiffness, $k_{bellows}$, is nonzero only when vacuum is applied; otherwise it is not connected to the vibration table during non-vacuum test conditions. An analysis of the model in Fig. 2.10 suggests that if $k_{bellows}$ is very small during high-vacuum testing, then the responses $x_2(t)$ and $x_3(t)$ should not change or will change very little between conditions (i) and (ii). This is achievable using the welded-bellows design of test rig A shown in Fig. 2.6. In contrast, applying the model of Fig. 2.10 to test rig B of Fig. 2.7 results in a rigid connection between the vibration table and drive assembly. The stiffness between the vibration table and drive assembly is much larger than that of test rig A due to the presence of the thick and heavy conflat flange joints used in rig B.

2.5.2 Accelerometer Response Comparison

A comparison between accelerometer responses for test rigs A and B is presented in Fig. 2.11 through Fig. 2.14. Figure 2.11 contains a baseline vibration signature of test rig A presented under high vacuum with the rotating assembly installed but not rotating. No attempt was made to further isolate the vacuum system from the drive assembly and base plate. Instead, the external inputs from the vacuum pumps, 60 Hz from the roughing pump and 1,250 Hz from the turbo pump, provided a traceable and known input to the system in addition to the input of highest interest, a 15 Hz excitation from the rotating assembly. The goal of the accelerometer tests

22   2 Vacuum Chamber Design

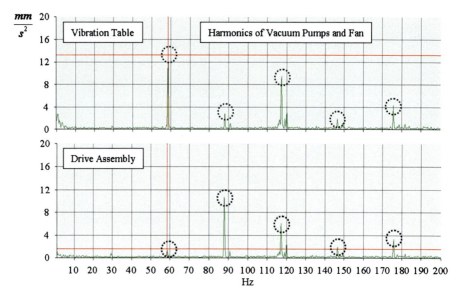

Fig. 2.11 Baseline response of the vibration table and drive assembly of test rig A under vacuum with the rotating assembly installed but not rotating (Reproduced with permission from Rev. Sci. Instrum. 82, 105113 Copyright 2011, AIP Publishing LLC)

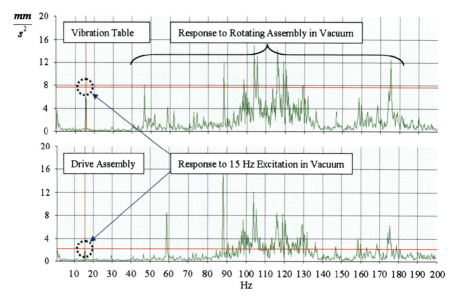

Fig. 2.12 Accelerometer response of the vibration table and drive assembly from test rig A. Response to a 15 Hz excitation from the rotating assembly at 8.6E-7 Torr (Reproduced with permission from Rev. Sci. Instrum. 82, 105113 Copyright 2011, AIP Publishing LLC)

2.5 Case Study: Chamber Vibration Isolation and Measurement

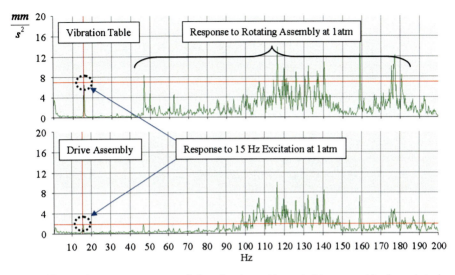

Fig. 2.13 Accelerometer response of the vibration table and drive assembly from test rig A. Response to a 15 Hz excitation from the rotating assembly at 760 Torr (Reproduced with permission from Rev. Sci. Instrum. 82, 105113 Copyright 2011, AIP Publishing LLC)

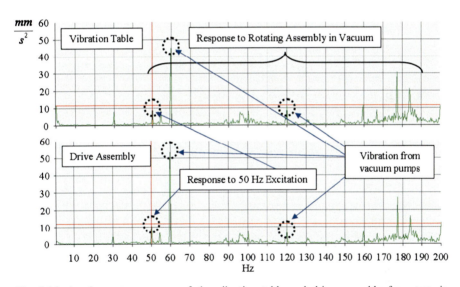

Fig. 2.14 Accelerometer response of the vibration table and drive assembly from test rig B. Response to a 50 Hz excitation from the rotating assembly at 3.6E-7 Torr (Reproduced with permission from Rev. Sci. Instrum. 82, 105113 Copyright 2011, AIP Publishing LLC)

was to check if the welded bellows would in any way alter the vibration response to the vibration table while the system is under vacuum. A cyclic force input is generated by applying an unbalance to the rotating assembly and then rotating it at 15 Hz. The nature of this force input is representative of the type of vibration characteristic to be detected under heat and vacuum conditions using these test rigs.

The accelerometer response of the vibration table and drive assembly under vacuum are presented in Fig. 2.12. Similarly, Fig. 2.13 has the accelerometer response of the vibration table and drive assembly at atmospheric pressure. Concerning Figs. 2.12 and 2.13, the values to compare are the acceleration magnitudes of the vibration table at 15 Hz under both vacuum and non-vacuum conditions. The acceleration responses of the vibration table and drive assembly for the vacuum and non-vacuum tests are nearly the same at 15 Hz. These results indicate that use of the welded bellows does not significantly influence the response of the vibration table while operating under vacuum. Specifically, the presence of the bellows with vacuum applied does not enable cross talk between the vibration table and drive assembly. For comparison, this result is equivalent to a very small spring constant, $k_{bellows}$, in Fig. 2.10 such that unwanted excitation from the drive assembly and its mass does not influence the response of the vibration table, $x_2(t)$. For comparison, the same rotating assembly was tested in test rig B and the results are presented in Fig. 2.14. Comparing the acceleration magnitude of Figs. 2.12 and 2.13 with that of Fig. 2.14 suggests difficulty in resolving low amplitude vibration. The 15 Hz input unbalance was not detectable in test rig B. Instead, the rotation speed of the rotating assembly was increased until the vibration magnitude response matched with that of test rig A. Specifically, the speed of the rotating assembly was increased to 50 Hz before reaching the magnitude response of test rig A. These results suggest that the mass and stiffness of the conflat flanges used in test rig B preclude sensitive and low amplitude vibration detection. Concerning test rig B in Fig. 2.7, it is impossible to isolate the vibration table from the drive assembly using thick-walled conflat-flange-type attachments. These results suggest that a welded-bellows assembly should be considered in the design optimization of the next test rig.

2.6 Case Study Continued: Optimum Chamber Design

Test rig C was built to accommodate larger diameter rotating assemblies and to enable operation at higher temperatures. This design incorporates conflat flange joints with the isolation aspects of the welded-bellows assembly used in test rig A. The new chamber, test rig C, is an extension of the chamber design of test rig B with inclusion of the welded bellows from test rig A. The singular assembly that was optimized for test rig C was the main chamber shown in Fig. 2.15. The chamber design has four aspects: two welded bellows, tubular interface attachments, tubing wall thickness, and chamber wall thickness.

As presented in Fig. 2.15, a conflat flange joint is used at one end of the chamber to enable removal and installation of the rotating assembly. During a test, the rotating assembly may be heated to 600 °C while the vibration response of the vibration table is measured over time. With the rotating assembly at 600 °C, the exterior of the chamber may reach 200 °C due to radiative heat transfer inside the chamber. Hard copper gaskets are required for all conflat joints due to

2.6 Case Study Continued: Optimum Chamber Design

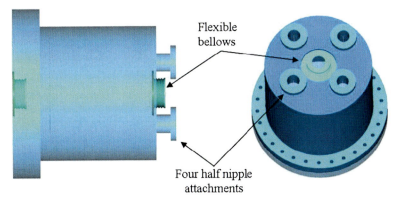

Fig. 2.15 Design optimization of chamber section of test rig C. The design couples the welded-bellows aspect with conflat flanges to enable vibration isolation at high temperature under vacuum (Reproduced with permission from Rev. Sci. Instrum. 82, 105113 Copyright 2011, AIP Publishing LLC)

the heating conditions of this test rig. The rotating assembly is driven by a hermetically sealed drive assembly similar to the one used on test rig A in Fig. 2.4. Two flexible bellows are used in the new design to enable more test configuration flexibility.

2.6.1 Design Optimization Analysis

The new chamber is made from 304 stainless steel (304SS), and all permanent seams are welded with no long-term gasket joints. Fabrication with stainless steel is expensive, and therefore, the amount of 304SS used should be minimized. The cost of cooling the chamber will be minimized through optimization of geometry and heat conduction through the half nipple attachments in Fig. 2.15. The interface attachment tube walls conduct heat from the chamber to the interface attachments. The design requirement concerning the half nipple attachments is that they cannot exceed room temperature, approximately 20 °C, to protect and isolate temperature-sensitive measurement equipment from heat damage and leakage. The conductive heat transfer through the attachment tubes will be minimized within the constraints that the tube section of the attachments be able to support a 25 N cantilever load with an end deflection less than 0.1 mm. This requirement applies equally to all four tubes so that any instrument or pump can attach at any interface flange.

The front surface of the chamber in Fig. 2.15 must not deflect more than 3 mm, and the chamber walls around the circumference must not deflect more than 1 mm. It is desirable that the entire chamber system has a first natural frequency greater than 600 Hz or three times the anticipated rotational speed of the rotating assembly. The cost of materials, fabrication, and operation to be minimized is given as

$$f(X) = \text{Cost}_{\text{metal}} + \text{Cost}_{\text{welding}} + \text{Cost}_{\text{operation}}. \tag{2.1}$$

The cost of the stainless steel for the chamber and tube interfaces is assumed constant. The cost of the stainless steel weld is assumed to be \$5/cm due to weld prepping of the surfaces and the required leak check after the weld is complete. The heat transferred to the chamber during the test emanates from the rotating assembly inside. Cooling costs per unit surface area of attachment tubes are based on use of one 120 VAC fan. Labor cost for assembling the test rig is not included in the optimization; rather, it is assumed that the test rig will be assembled by a staff engineer or laboratory technician with basic knowledge of hand and shop tools. The labor costs associated with assembling the rig are included in the overhead costs of the facility.

The design constraints for the problem will first be stated and then translated into mathematical form using established equations and expressions found in Young (1989), O'Hanlon (1989), and Incropera and DeWitt (1996). The design constraints are:

1. Chamber diameter is fixed at 36 cm.
2. Length of the chamber is at least 30 cm.
3. Chamber wall thickness strong enough to maintain pressure of 10^{-8} Torr.
4. The chamber first natural frequency must be greater than 600 Hz.
5. Force and displacement transmissibility of rotating part attachments less than 0.1.
6. Deflection of the front plate less than 0.1 mm but should be as small as possible.
7. Diameter and length of attachments on front plate must satisfy minimum molecular flow conductance requirement for a 60 L/s pump. Pumps and instrumentation are allowed to mount on any of the four attachments.
8. Wall thickness and length of the front plate attachments must not conduct heat to half nipple flange interfaces.
9. The attachments on the front plate must be capable of supporting up to a 25 N load, the nominal weight of instrumentation such as pyrometers, analyzers, and laser probes.
10. The first natural frequency of the attachments must be greater than or equal to that of the chamber, 600 Hz.

Constraints 1 and 2 are classified as size constraints and 3 through 10 may be classified as material and behavioral constraints. Expressions and references for material and behavioral constraints 3 through 10 are established in the fields of mechanical engineering and applied physics and are presented below. The mathematical form of the optimization problem is stated as,

$$\min f(X) = \rho(A_1 L_1 + A_3 L_3 + A_2 t_2)\text{Cost}_{\text{material}} + \pi(2d_1 + 8d_3)\text{Cost}_{\text{weld}} + h \\ \times \text{Cost}_{\text{cooling}} \tag{2.2}$$

subject to the constraint expressions in Eqs. 2.3, 2.4, 2.5, 2.6, 2.7, 2.8, 2.9, 2.10, 2.11, 2.12, 2.13, 2.14, 2.15, and 2.16 for the chamber shown in Fig. 2.15. Constraints g_1 through g_3 relate to radial and hoop stress and deflection of the chamber:

2.6 Case Study Continued: Optimum Chamber Design

$$g_1 = p\frac{d_1}{t_1} - \sigma_y \leq 0 \tag{2.3}$$

$$g_2 = \frac{p\left(\frac{d_1}{2}\right)^2}{Et_1}\left(1 - \frac{\nu}{2}\right) - \text{cdif} \leq 0 \tag{2.4}$$

$$g_3 = p\frac{d_1}{2t_1} - \sigma_y \leq 0. \tag{2.5}$$

Constraints g_4 and g_9 account for the first bending mode natural frequency of the chamber section and for each attachment tube, respectively. Since the rotating device inside the chamber may have an unbalance, or could develop an unbalance during testing, it is good practice to design for at least 2× the frequency response of the chamber system to internal excitation:

$$g_4 = \text{freq} - \frac{6.93}{2\pi}\sqrt{\frac{EI_1 g}{wL_1^3}} \leq 0 \tag{2.6}$$

$$g_9 = \text{freq} - \frac{1.73}{2\pi}\sqrt{\frac{EI_3 g}{wL_3^3}} \leq 0. \tag{2.7}$$

Constraints g_5 and g_6 relate to front plate stress and bending deflection. With the chamber evacuated, under vacuum, the external pressure on the chamber is 14.6 psi. The load on the chamber is opposite that of a pressure vessel. For the present case study, the chamber must not deflect during heating:

$$g_5 = \frac{p\left(\frac{d_1}{2}\right)^4}{64D}\frac{(5+\nu)}{(1+\nu)} - \text{pdif} \leq 0 \tag{2.8}$$

$$g_6 = 6\frac{\left(\frac{d_1}{2}\right)^2}{16}\frac{(3+\nu)}{t_2^2} - \sigma_y \leq 0. \tag{2.9}$$

Limits on bending and deflection of instrumentation attachments is accounted for using constraints g_7 and g_8. The attachments need to support the weight of instruments as well as maintain vacuum integrity for the life of the test:

$$g_7 = \frac{2wL_3^2}{6EI_3} - \text{adif} \leq 0 \tag{2.10}$$

$$g_8 = \frac{wL_3}{I_3}\left(\frac{d_3}{2} - t_3\right) - \sigma_y \leq 0. \tag{2.11}$$

Constraints g_{10}, g_{11}, and g_{12} relate to hoop stress and radial wall deflection in the half nipple attachment tubes. For some test applications, the wall thickness of an attachment tube may need to be minimized to allow optimal passing of magnetic fields through the wall. For example, if electron-beam heating through one of the half nipple attachments is needed, then external magnets are used to guide the electrons through nipple. The magnets are typically attached to the exterior of the tube so the magnetic fields must pass through the tube wall:

$$g_{10} = p\frac{d_3}{t_3} - \sigma_y \leq 0 \tag{2.12}$$

$$g_{11} = p\frac{d_3}{2t_3} - \sigma_y \leq 0 \tag{2.13}$$

$$g_{12} = \frac{p\left(\frac{d_3}{2}\right)^2}{Et_3}\left(1 - \frac{v}{2}\right) - adif \leq 0. \tag{2.14}$$

The constraints g_{13} and g_{14} track molecular flow and heat transfer in the attachment tubes. Molecular flow through the attachment tubes will determine system pump-down time and may limit recovery during chamber outgassing. In fact, there may be a decade difference in pressure between the pump and the chamber if the attachment tube is too small. In addition, the larger diameter tube attachment will increase convective heat flow to the ambient air and will assist with chamber cooling:

$$g_{13} = \text{pump} - 7.66 \times 10^5 \frac{\pi d_3^2}{4(1-pr)} pr^{0.712}\sqrt{1-pr^{0.288}} \leq 0 \tag{2.15}$$

$$g_{14} = T_2 + (T_1 - T_2) \times \exp\left(-L_3\sqrt{\frac{h \times pc}{k \times ac}}\right) - T_2 \leq 0. \tag{2.16}$$

Equations 2.3, 2.4, 2.5, 2.6, 2.7, 2.8, 2.9, 2.10, 2.11, 2.12, 2.13, 2.14, 2.15, and 2.16 may be found in Young (1989), O'Hanlon (1989), and Incropera and DeWitt (1996). While optimizing the chamber parameters, be sure that you can justify your optimization analysis with equations from a good reference to take advantage of previous work of others to improve the design of your chamber.

The calculated variables in Eqs. 2.3, 2.4, 2.5, 2.6, 2.7, 2.8, 2.9, 2.10, 2.11, 2.12, 2.13, and 2.14 are defined as

$$I_1 = \pi\left(\frac{d_1}{2} - d_1\right)^3 t_1, I_3 = \pi\left(\frac{d_3}{2} - d_3\right)^3 t_3, D = \frac{Et_2^3}{12(1-v^2)}, \tag{2.17}$$

$$ac = \pi\left(\frac{d_3}{2} + t_3\right)^2 - \pi\left(\frac{d_3}{2} - t_3\right)^2, pc = \pi d_3, A_1 = \frac{\pi}{4}\left[(d_1 + 2t_1)^2 - d_1^2\right], \tag{2.18}$$

$$A_3 = \frac{\pi}{4}\left[(d_3 + 2t_3)^2 - d_3^2\right], A_2 = \frac{\pi}{4}d_1^2 - 4\pi d_3^2. \tag{2.19}$$

The associated design vector, $X = \{L_1, t_1, t_2, L_3, t_3, d_3, h\}$ for Eq. 2.2, contains seven independent design variables. The design variables are defined as L_1 and t_1, the

2.6 Case Study Continued: Optimum Chamber Design

length and wall thickness of the chamber, and t_2, the thickness of front plate section. The variables L_3, t_3, and d_3 are the length, wall thickness, and diameter of the attachment tubes, respectively. The heat transfer coefficient h applies to the half nipple tube attachments only. The bounds on the design variables were chosen with observation of readily available flanges, nipples, and tubes. Based on a survey of off-the-shelf components from leading vacuum hardware retailers, the bounds on the design variables were chosen as

$$300 \leq L_1 \leq 390, \ 1.5 \leq t_1 \leq 5, \ 1.5 \leq t_2 \leq 5, \ 50 \leq L_3 \leq 150, \\ 1.5 \leq t_3 \leq 5, \ 30 \leq d_3 \leq 70, \ 80 \leq h \leq 150. \quad (2.20)$$

All lengths in Eq. 2.20 have units of mm, and h has units of $\frac{W}{m^2 \cdot K}$. The bounds on the thicknesses t_1, t_2, and t_3 were chosen with respect to weldability and manufacturability range of material thicknesses for 304 stainless steel tubing and flat stock. For this analysis, the following fixed material and system parameters were used:

$$d_1 = 350\,\text{mm},\ g = 9.81\frac{\text{m}}{\text{s}^2},\ k = 14\frac{\text{W}}{\text{m}\cdot\text{K}},\ w = 25\,\text{N},\ E = 210\,\text{GPa},\ \nu = 0.29 \\ \sigma_y = 280\,\text{MPa},\ p = 100.6\,\text{kPa},\ \text{freq} = 600\,\text{Hz},\ \text{pump} = 60\frac{\text{L}}{\text{s}},\ pr = 0.52, \\ T_1 = 200°\text{C},\ T_2 = 20°\text{C},\ cdif = 1\,\text{mm},\ pdif = 3\,\text{mm},\ adif = 0.1\,\text{mm}. \quad (2.21)$$

The parameter pr is the ratio of the vacuum pump pressure to chamber pressure. The parameters $cdif$, $pdif$, and $adif$ are the chamber radial deflection, end plate deflection, and attachment deflection limits, respectively. The parameter p in Eqs. 2.3, 2.4, 2.5, 2.6, 2.7, 2.8, 2.9, 2.10, 2.11, 2.12, 2.13, 2.14, 2.15, 2.16, 2.17, 2.18, 2.19, 2.20, and 2.21 accounts for atmospheric pressure at sea level in kilopascals applied to the external surfaces of the chamber and nipple sections. The cost parameters in the objective function of Eq. 2.2 are defined as

$$\text{Cost}_{\text{material}} = \$2.39/\text{kg},\ \text{Cost}_{\text{weld}} = \$500/\text{m},\ \text{and}\ \text{Cost}_{\text{cooling}} = \$7.52/\text{h}. \quad (2.22)$$

The cooling costs for this rig are a function of the electricity used to operate the cooling fan. Defining h in Eq. 2.20 as a design variable allows flexibility within the optimization problem to reach the lowest cost cooling solution using fans. The program does not consider the details of the heat transfer coefficient, such as surface area, heat flow, and temperature differential, etc. Rather, the program considers the amount of h that is needed to meet the design constraints. The cost per unit h was calculated based on 8,000 h of testing at an electricity cost rate of $\$0.877/\text{kWh}$.

The optimization problem presented in Eqs. 2.2, 2.3, 2.4, 2.5, 2.6, 2.7, 2.8, 2.9, 2.10, 2.11, 2.12, 2.13, 2.14, 2.15, 2.16, 2.17, 2.18, 2.19, 2.20, and 2.21 contains nonlinear constraints and is therefore classified as a nonlinear programming problem. This problem is solved using a sequential quadratic programming algorithm found in Rao (2009). The basis of quadratic programming algorithms is to solve the

Table 2.1 Results for three optimization options for new system, test rig C

	1. fval = $1049.20		2. fval = $837.81		3. fval = $983.74	
	Upper bound	Medium fan Freq. minimum 3× run speed	Upper bound	Small fan Freq. minimum 3× run speed	Upper bound	Medium fan Freq. minimum 2× run speed
L_1	390 mm	300 mm	Unchanged	300 mm	Unchanged	300 mm
t_1	5 mm	1.5 mm	Unchanged	1.5 mm	Unchanged	1.5 mm
t_2	5 mm	4.72 mm	Unchanged	4.72 mm	Unchanged	4.72 mm
L_3	150 mm	145.52 mm	170 mm	170 mm	Unchanged	150 mm
t_3	5 mm	1.5 mm	Unchanged	1.5 mm	Unchanged	1.5 mm
d_3	70 mm	70 mm	90 mm	81.28 mm	Unchanged	55.71 mm
h	150 W/m^{2}K	111.43 W/m^{2}K	Unchanged	81.65 W/m^{2}K	Unchanged	104.87 W/m^{2}K

Reproduced with permission from Rev. Sci. Instrum. 82, 105113 Copyright 2011, AIP Publishing LLC

Lagrange multiplier problem directly with application of the Kuhn-Tucker conditions for optimality. The optimization calculations were carried out in **MATLAB** using the *fmincon* function in the Optimization Toolbox. The *fmincon* function requires that the system of constraint equations be continuous and differentiable within the upper and lower bounds of the constraint variables. At each iteration, the Kuhn-Tucker conditions are solved through calculation of the second order Hessian matrix. If all equations are quadratic, a solution will be reached within two iterations. Due to the non-quadratic nature of the constraint equations, numerical Hessian updating is needed, and the optimum solution was located after 12 iterations.

2.6.2 Compare Optimum Designs

Results of the top three optimization runs are presented in Table 2.1 as Options 1, 2, and 3. Option 1 is the optimum solution using the initial upper and lower bound constraints on the design vector as given by Eq. 2.20. Option 2 is the optimum solution using a slightly modified upper bound vector, as indicated in Table 2.1. Option 3 is the optimum solution using the same upper and lower bounds as Option 1, but with a chamber first natural frequency limit of 400 Hz instead of 600 Hz.

An inspection of the program output for Option 1 indicates that the inequality constraints 5, 9, and 14 are active at the final design. For example, constraint 5 influences the front plate thickness, and constraints 9 and 14 control the attachment tube first frequency and the half nipple attachment convection coefficient, h, respectively. Constraints 4 and 13 are very far from influencing the final design as was indicated by their large negative values from the program output compared with other constraints in the program output.

2.7 Assembly and Cleaning

The final cost of fabricating the chamber parts and operating the new test rig is mostly influenced by the attachment tube geometries. The cost of off-the-shelf flanges and other parts common to both test rigs B and C have been excluded from the results in Table 2.1. The cost of common parts for all designs is about $2,000.00. Two competing constraints on the attachment half nipple structure were stiffness and heat conduction in the tube walls. For example, increasing the wall thickness will increase the first natural frequency, but it will also increase the heat conduction from the chamber and ultimately require more forced convection cooling which will drive up the cost.

The cooling costs associated with test rig B when testing a rotating assembly at 600 °C were compared with the results in Table 2.1 for the new rig, test rig C. The cooling fan costs alone for test rig B were more than double the cooling costs of the options presented in Table 2.1. Test rig B requires two large cooling fans during operation in order to meet the 20 °C temperature constraint on the half nipple attachments. The cooling cost of test rig B is $15.04/h for comparable ranges of h presented in Eq. 2.20 due to the necessity of the two large fans instead of one medium-sized fan for test rig C. The operating costs of the designs presented in Table 2.1 are about half the operating cost of test rig B based on 8,000 h of testing.

Referring to Option 2 in Table 2.1, the program reached a compromise by increasing the half nipple attachment tube diameter and length and effectively raising the first natural frequency while only slightly raising tube heat conduction and length and improving forced convection efficiency via increased surface area. However, this option is problematic in that larger attachment tube diameters could result in accessibility issues among measurement instrument attachments. Option 2, though least expensive to build and operate, could restrict measurement hardware attachment accessibility and necessitate custom attachment interfaces on the instruments. Option 3 is less expensive than Option 1, but it also has a lower first natural frequency capability, and the diameter of the half nipple attachments is close to the minimum desirable. The reduced frequency capability of Option 3 may render the rig obsolete. In comparing all three options, Option 1 meets all the upper bound requirements as outlined in Eq. 2.20 without any modifications.

2.7 Assembly and Cleaning

The manufacturing history of all vacuum components can have a profound influence on the process results. For example, if the welded joints within the chamber were fabricated using dirty tools, including trace elements of cutting or grinding tool metals, oils, paint, lint from clothing or clothes, etc., then the chamber could outgas for months, thereby corrupting the vacuum space inside. As a best practice, only lint-free towels and high purity alcohol should be used to wipe down the inside of the chamber and components. Oil-free gloves and hairnets should be worn at all times when the chamber is open. One piece of hair, glove, or even finger prints can cause excessive pressure in the chamber, which will delay the process during pump down.

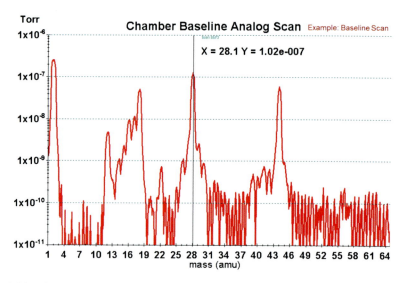

Fig. 2.16 RGA scan of chamber with oxygen (16 amu) and nitrogen (28 amu) present, and hydrocarbon or oil contamination (44 amu)

2.8 Residual Gas Analysis

Residual gas analysis (RGA) is an excellent chamber diagnostic tool for commissioning a new system. RGA should also be used after any major work has been done to the chamber or its components. For example, during a complete system teardown and reassembly, there are lots of opportunities for contamination. Conducting an RGA scan and then comparing it to the baseline scan, when the chamber was new, will put to rest any suspicions about the cleanliness, or it will at least confirm that the chamber is still contaminated.

The SRS100 RGA system is used in Chaps. 6 and 7 to track exiting argon gas from the coating layers during the ion-plating process. Using RGA, one can track the concentration of argon present in the system to confirm correct operation of the argon mass-flow controllers. The vacuum chamber for the rolling contact fatigue (RCF) test rig in Chap. 6 uses RGA to assess outgassing during run-in of the silver-coated balls. The RGA tracks the amount of argon that has been squeezed out from the silver coating during run-in. When all of the argon is gone, the silver on the surface of the balls has been sufficiently run-in.

2.8.1 Establishing Chamber Baseline

A residual gas analysis of a chamber similar to Figs. 2.9 and 2.15 is presented in Fig. 2.16. The data is presented as a frequency spectrum of mass elements in atomic mass units (amu). With this spectrum, one can deduce what gases are present in the chamber under vacuum. For example, there are three peaks of interest in Fig. 2.16, and the 28 amu peak indicates that nitrogen N_2 is present in the chamber. The other two peaks of interest indicate that there is oxygen present as well, $O_2 \rightarrow$ 16 amu, and some form of oil contamination showing up as a hydrocarbon molecule at $C_nH_n \rightarrow$ 44 amu. If the process is sensitive to nitrogen, oxygen, or oil, the presence of these contaminant molecules and gases in sufficient quantity would corrupt and alter the deposition process. If the hydrocarbon molecules are present during the RCF test, they could influence the fatigue life of the silver coating. The amount of each gas present correlates with the partial pressure of the chamber. Partial pressure is indicated by the height of each peak given in units of Torr. The sum of all peaks is the total pressure inside the chamber.

2.8.2 Helium Leak Check

Prior to operation of the chamber, a leak check using helium gas should be done to confirm that there are no slow leaks in the system. Helium gas is used because it is the second smallest molecule on the periodic table, second to hydrogen. Literally, the small helium molecule and/or atoms will seep through any gaps within the chamber structure. It is important to detect and repair leaks in the system as soon as possible. For example, even if the chamber reaches "good" vacuum, to 10^{-6} Torr range, a very small leak in the chamber will allow contaminant gas to corrupt the process inside, even though the base chamber pressure is sufficiently low. Large pumping capacity, for example, will hide the presence of a leak. This is of particular concern with cryogenic systems which would require more frequent regeneration if a small leak was present, but would maintain "good" vacuum regardless.

2.9 System Commissioning

After assembly, cleaning, leak check, and RGA baseline, the system is ready for commissioning. Vacuum systems are not all the same, and slight differences in control connections, grounding, and cooling can make each system unique with respect to process recipe. Documentation, operating instructions, and data logs should be reviewed before the system can be trusted for use. More detailed best practices may be found in Mattox (1998); the reader is encouraged to consult this text as well when designing and commissioning any vacuum chamber application.

References

Baglin V, Collins I, Grobner O. Photoelectron yield and photon reflectivity from candidate LHC vacuum chamber with implications to the vacuum chamber design. LHC project report 206. Stockholm: EPAC 1998 Conference Proceedings; 1998. p. 1–4.

Iijima N. Structure for vibration isolation in an apparatus with a vacuum system. United States patent 4,352,643, 5 Oct 1982.

Incropera F, DeWitt D. Fundamentals of heat and mass transfer. 4th ed. New York: Wiley; 1996.

Lee C, Wang D, Choi W. Design and construction of a compact vacuum furnace for scientific research. Rev Sci Instrum (AIP). 2006;77(12):1–5.

Mattox D. Handbook of physical vapor deposition processing. Westwood: Noyes; 1998.

O'Hanlon JO. User's guide to vacuum technology. 2nd ed. New York: Wiley; 1989.

Rao S. Engineering optimization. 4th ed. New York: Wiley; 2009.

Young W. Roark's formula for stress and strain. New York: McGraw-Hill; 1989.

Chapter 3
Rolling Contact Testing of Ball Bearing Elements

Abbreviations

W	Applied contact load
ψ	Chemical potential in thermodynamic analyses
W_Y, F_Y	Contact load at material yield
Γ	Contact loading-type constant
H	Contact material hardness (GPa)
U_{ab}, U	Contact surface energy elastic energy material
β	Cup surface contact angle
$S - N$	Cycles versus load
d_A	Diameter of debris particle from friction contact
δ	Differential operator $\frac{\partial}{\partial q_n}$
K_D	Effective contact radius for contact stress calculation
E^*	Effective modulus
R	Effective radius of curvature
p_o	Hertz contact pressure
a_r	Hertz contact radius
c, d, η	Load constants for contact stress calculation
p_m	Mean applied contact pressure
N_c, N_r	Normal force between Cup–ball and ball–rod
$\nu_{1,2}$	Poisson's ratio of contact materials
σ_i	Principle stress
$R_{1,2,3,4}$	Radius of curvature of balls and rod
$\dot{S}$	Rate of change entropy
D_r	Rod diameter
J_2	Second stress invariant
$E_{1,2}$	Young's modulus of contact materials
CVFC	Control volume fraction coverage

© Springer International Publishing Switzerland 2015
M. Danyluk, A. Dhingra, *Rolling Contact Fatigue in a Vacuum*,
DOI 10.1007/978-3-319-11930-4_3

RCF   Rolling contact fatigue
TiN   Titanium nitride

Rolling contact testing requires understanding of friction, lubrication transfer, and a common framework from which to calculate contact stress between rolling contact elements. So long as the calculations and assumptions for each of these topics are consistently applied across all experiments, the resulting experimental data is very useful to assess coating performance. Often, data from tests involving friction are fit to an empirical model set of equations. There are several references concerning the foundational aspects of friction and wear, "Friction and Wear of Materials," by Rabinowicz, and "Engineering Tribology," by Stachowiak and Batchelor (2005) to name a few. More recently, textbooks concerning wear and friction of thin film coatings and surface engineering have emerged as well, "Coatings Tribology: Properties, Techniques and Applications in Surface Engineering," by Holmberg and Matthews (2009), and "Surface Modification and Mechanisms," by Totten and Liang. A more general approach to friction has been proposed by Nosonovsky and Mortazavi (2013), considering friction and the associated processes of wear as a universal and general phenomenon, independent of how the friction is generated. This approach removes the distinction between wear in rolling contact systems, such as ball bearing sets and sliding contact wear mechanisms in reciprocating machinery, for example, and presents a thermodynamic connection for all types of wear.

The focus of this chapter is to review the assumptions and calculations used to calculate contact stress in rolling contact fatigue (RCF) testing. The goal of this chapter is to establish a framework from which thin film coatings applied to ball bearings tested in rolling contact fatigue and in high vacuum may be evaluated and compared across variation in coating process parameters. The calculations presented in this chapter are used in Chaps. 4 and 6 to help quantify the effects of process aberration on coating RCF life.

3.1 Introduction

For elements in rolling contact, surface damage occurs due to cyclic stress loading between the contacting surfaces. Though not immediately obvious, cyclic stress loading at a contact surface results in subsurface crack initiation and is considered a form of surface wear. Fatigue wear is the primary failure mode of rolling element contacts for both liquid film and solid film lubrication systems. Fatigue wear is the weakening of the contact surfaces and the material immediately below it due to cyclic loading. The specific mechanism of fatigue wear is accumulation of plastic strain beneath the contact surface over several million cycles depending on the applied contact load. For a more detailed development of these topics, the reader is encouraged to review Bhushan (1999) and Rabinowicz (1995).

3.1 Introduction

Damage due to fatigue wear may be reduced in three ways: (i) minimizing internal stress-raising material defects, (ii) improving surface finish and smoothness of the contacting surfaces, and (iii) reducing shear stress at the surface contact. Concerning elements in rolling contact such as ball bearings, the first two of these methods are directly related to material processing and post material handling. The third point, reducing shear stress at the contact surface, may be achieved through application of thin film lubrication as suggested by Bhushan (1999) and Mattox (1998). But even with thin film lubrication, after many cycles and depending on the magnitude of the normal load, damage will accumulate in the form of plastic deformation within the lattice structure of the substrate ball material. Eventually a maximum shear stress condition results and a small internal crack is initiated within the material below the point of contact.

If the lubricating film on the ball becomes depleted or is otherwise ineffective, the contacting surfaces begin an abrasive-wear condition. The contacting surfaces between the ball and race begin to remove material from one another due to increased friction, brought on from lack of lubrication. The material transfer is similar to a closed-contact three-body abrasive-wear situation. The particles generated from the contacts are free to move between the surfaces, but they do not adhere to either surface. In this situation, the total number of particles increases as more material is removed from each surface. Subsurface crack growth increases as well due to the increased friction on the surface associated with the abrasive-wear condition. When the film is sufficient for lubrication, the abrasive-wear condition is reduced. Particles are still being generated but at a lower rate. If there is little or no flowing medium to remove the particles, such as oil or flowing gas, the contacting surfaces favor an adhesive-wear situation such that the particles are redistributed and adhere to each contacting surface. Liquid flow lubrication provides a way to constantly remove debris from the contact area so that adhesive wear is reduced. Both of these conditions, abrasive and adhesive wear, increase shear stress within the contact materials and increase the likelihood of subsurface crack initiation. Concerning rolling contact fatigue (RCF) testing, fatigue wear is the dominant mechanism in that the alternating stress on the surface causes crack initiation and subsequent growth until the crack reaches the surface and causes a spall.

Wear modeling and prediction of rolling contact ball bearings in vacuum is very difficult. While it is possible to identify the transition between individual mechanisms of the wear, i.e., abrasive wear, adhesive wear, and fatigue wear, it is difficult to identify the exact mechanism without halting the test for autopsy. For example, work hardening of the surfaces in rolling contact will start during the beginning of the test process and will continue through the length of the test. The fatigue-wear mechanism dominates during most of an RCF test, and as the lubricating film is being depleted, one can detect a change in vibration of the test corresponding to changes in the film and subsurface material. The difficulty is that if the ball and raceway surfaces have been work hardened over many cycles, then the magnitude of the frequencies associated with vibrations near the end of the test will differ from those observed at the beginning of the test due to work hardening alone. Changes in vibration response will also occur due to transition between wear mechanisms, and

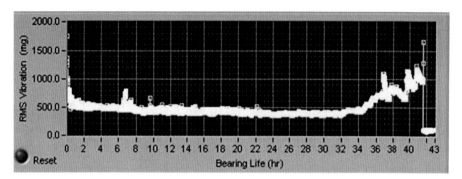

Fig. 3.1 Vibration response over time of a rolling contact fatigue test using solid film lubrication in high vacuum

it is not always obvious what mechanism or surface change caused the vibration response to change.

Vibration change over time is illustrated in Fig. 3.1 for a RCF test using nickel–copper–silver film lubrication. The test was run at 8,400 rpm in high vacuum using six silver-coated ball bearings at 2.1 GPa contact stress. The high vibration within the first hour of the test is the so-called run-in period for the contacting surfaces along with the onset of work hardening. For most of the test, from 2 to 34 h, a fatigue-wear condition dominates. There is a trend for decreasing vibration over the length of the test and then an abrupt increase signifying a change in wear mechanism. The vibration change after 34 h is the start of an abrasive-wear condition as the film lubrication became depleted. The spikes in vibration between 6 and 8 h, and similarly between 36 and 38, are brief examples of adhesive wear within the solid lubrication system. For the results presented in Fig. 3.1, the brief spikes occurred due to film flaking and redistribution on the surface. The test was stopped at 41 h due to excessive vibration and subsequent autopsy revealed total depletion of the film lubricant from the surfaces of all the balls. More RCF test results will be presented and discussed in Chaps. 4 and 6.

Understanding wear and friction requires knowledge of the type of contact between the two solid bodies. Incipient sliding occurs when two contacting bodies are pressed together such that a stick point exists within the contact area and materials from each body slide relative to each other about that point. There is no imposed relative motion between the bodies during incipient sliding condition, rather sliding occurs due to elastic deformation of each surface. One of the tribological benefits of coatings such as silver, lead, and MoS_2 is that their compliant nature reduces subsurface stresses during incipient sliding. Of these three films, silver in particular has the best load carrying capability and RCF life when operating at 400 °C in high vacuum. Silver lubrication on ball bearings has been used extensively in high-voltage and high-vacuum devices such as rotating anode x-ray tubes. Silver as a lubricant is nonreactive in the high-voltage and vacuum environment inside an x-ray tube. Silver is has low shear strength which helps to

3.2 Analysis of Rolling Contact in Vacuum

reduce friction during rolling contact. For more information about the tribological aspects of silver, and other lubricant films such as MoS_2 and lead, refer to "Coatings Tribology," by Holmberg and Matthews (2009).

3.2 Analysis of Rolling Contact in Vacuum

Solid lubricants find unique application in systems where the pressure range is 10^{-3}–10^{-8} Torr, and temperature may range between -100 and $500\,°C$. As far back as 1972, NASA Report SP-5059, SP-5059(01) (1972) outlined testing platforms for high-vacuum and high temperature testing involving sliding contact specimens, rolling contact ball bearings, metal disk and riders, and sliding contact of a sleeve on a cylindrical rod. Lubricant film survival is highly dependent on adhesion and substrate subsurface bonding.

3.2.1 Calculations of Contact Stress

A calculation of contact stresses in rolling elements was first established by Hertz in 1882. The calculated stress due to ball–ball or ball–rod contact is referred to as Hertz stress and is typically 1.5 times greater than the applied pressure of the contacts if both surfaces were flat. For two spheres with radii R_1 and R_2, the effective radius of contact R is used in the stress calculation as

$$\frac{1}{R} = \frac{1}{R_1} + \frac{1}{R_2}. \qquad (3.1)$$

The contact radius a_r accounts for the contact area and is highly dependent on material properties and the shape of the contacting elements. The Hertz contact radius is defined as

$$a_r = \frac{\pi p_0 R}{2E^*}, \qquad (3.2)$$

where p_0 and E^* are the contact pressure and effective modulus of the contacting materials. The effective modulus used for calculation of contact stress includes the Poisson ratio of each material as

$$\frac{1}{E^*} = \frac{1-\nu_1^2}{E_1} + \frac{1-\nu_2^2}{E_2}. \qquad (3.3)$$

The contact pressure p_0 is always larger than the mean applied pressure such that $p_0 \simeq 1.5 p_m$. The contact pressure is related to the contact area as

$$p_0 = \frac{3W}{2\pi a_r^2}, \tag{3.4}$$

with W defined as the applied load to the contacts. The calculations presented forthwith have been derived in Bhushan (1999) and are reproduced here for clarity. The load required to initiate yield in either contacting material is defined as

$$W_Y = 21.17R^2 Y \left(\frac{Y}{E^*}\right)^2 \tag{3.5}$$

with the variable Y related to material hardness, H, and given as $H \approx 2.8Y$. The depth of maximum shear stress in either material has been shown to be $d = 0.48a_r$. This depth has been derived using the von Mises shear strain criterion and the second stress invariant

$$J_2 = \frac{1}{2}S_{ij}S_{ij} \equiv \frac{1}{6}\left\{(\sigma_1 - \sigma_2)^2 + (\sigma_2 - \sigma_3)^2 + (\sigma_3 - \sigma_1)^2\right\}, \tag{3.6}$$

where σ_1, σ_2, and σ_3 denote principal stresses in the contact material. Presented next is a simple example involving contact area calculation for two spheres in contact.

Example 3.1: Contact Stress Between Two Spheres Consider a contact analysis between two spheres: one ceramic 9.53 mm diameter with properties $E_{Cer} = 430$ GPa, $\nu_{Cer} = 0.3$, and $H_{Cer} = 17$ GPa and the other 12.7 mm diameter steel ball with properties $E_{Stl} = 200$ GPa, $\nu_{Stl} = 0.29$, and $H_{Stl} = 4$ GPa. Calculate the normal force required to yield the balls.

Solution: Since the steel ball has a lower hardness than the ceramic ball, we can expect yielding to occur first in the steel ball. To calculate the normal force required to yield the steel ball, we first calculate the composite modulus E^* using Eq. 3.3 as

$$E^* = \left(\frac{1 - 0.3^2}{430} + \frac{1 - 0.29^2}{200}\right)^{-1} = 149.3 \text{ GPa}. \tag{3.7}$$

Calculate the composite radius R using Eq. 3.1, as

$$R = \left(\frac{1}{4.75} + \frac{1}{6.35}\right)^{-1} = 2.72 \text{ mm}. \tag{3.8}$$

Using Eq. 3.5, yielding in the steel ball is expected to occur at force F_y as

$$F_y = 21.1 \times (0.00272)^2 \frac{\left(\frac{4 \times 10^9}{2.8}\right)^3}{\left(149.3 \times 10^9\right)^2} = 20.42 \text{ N}. \tag{3.9}$$

The radius of the contact area is calculated using Eq. 3.2 or

3.2 Analysis of Rolling Contact in Vacuum

Fig. 3.2 Force balance of cup–ball–rod contact

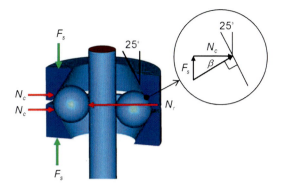

$$a_r = \left(\frac{3 \times 20.42 \times 0.00272}{4 \times 149.3 \times 10^9}\right)^{\frac{1}{3}} = 0.065 \text{ mm}. \quad (3.10)$$

With a contact radius of 0.065 mm and using $d = 0.48\, a_r$, we can expect yield will occur 0.031 mm below the surface of the steel ball. Crack initiation and subsequent spall will start 0.031 mm below the contact surface.

Contact stress is a calculated quantity and is therefore dependent on the assumptions made at the start of the analysis. In some cases, it may be better to present RCF test results as load versus cycles rather than stress versus cycles. However, some design applications are based on stress analyses so the calculation of contact stress of the rolling elements in contact may be needed for material selection. The first step is to establish a systematic approach to the calculation process for the RCF load condition. Consider the force balance in Fig. 3.2. The load F_s is applied to the cups in the vertical direction, usually using coil springs if the test is run inside a vacuum chamber. The cup angle $\beta = 25°$ is used to calculate the normal load N_c between the cup and ball. The relation between F_s and N_c is

$$N_c = \frac{F_s}{\tan(\beta)}. \quad (3.11)$$

The resultant normal load N_r applied to the rod is twice the load applied between the ball and cup or $N_r = 2N_c$. The contact stress between the ball and rod is a useful design parameter to assess a candidate material in rolling contact fatigue. For the RCF platform, the test material and the stress of interest is between the ball and rod. The stress between the ball and cup may be calculated as well, if needed.

A systematic calculation of contact stress is found in Chap. 13 of Young (1989). The data and analyses methods in Young (1989) grew out of the rail road industry in the early nineteenth century, sponsored in part by organizations such as The American Railway Engineering Association. Calculation of stress for various shapes of contacting elements is presented in Table 33 of Young (1989) and, specifically, contact between a sphere and rolling element such as the ball and rod configuration used in the RCF method. The stress between contacting curved elements is not always linearly proportional to the normal contact load N_r. In fact, due to the very small contact area between curved surfaces, the contact stress is typically very large compared to N_r even for light contact loads. For example, the calculation in Example 3.1 using Eqs. 3.8, 3.9, and 3.10 is most influenced by the radius of the smaller diameter ball. The contact between the ball and rod of the RCF test is designed to maximize contact stress for the minimum applied contact load N_r. In fact, for the cup–ball–rod test configuration, it is possible to increase the contact stress on the rod, the candidate test material, simply by reducing the rod diameter for the same ball size and applied contact load N_r.

Contact stress between the ball and rod may be calculated using the general case for two body contact with curvatures. Consider each contact element to have two radii of curvature, in each plane 90° apart. The curvatures are designated as R_1 and R_2 for element 1, and R_3 and R_4 for element 2. If we make element 1 the ball, then $R_1 = R_2$. We can define element 2 as the rod with diameter of $D_r = 2R_3$, and then let the radii $R_4 \to \infty$, effectively making it a planar surface in the axial direction of the rod. Following the analysis of Young (1989) in Chap. 13, the contact stress is calculated as

$$\sigma_{rc} = \frac{1.5 N_r}{c \pi d}, \qquad (3.12)$$

where the constants c and d are defined using Young variables as

$$c = \alpha \sqrt[3]{\frac{N_r K_D}{E^*}}, \qquad (3.13)$$

$$d = \beta \sqrt[3]{\frac{N_r K_D}{E^*}}. \qquad (3.14)$$

The composite modulus E^* has been defined in Eq. 3.3. The constant K_D is defined as

$$K_D = \frac{1.5}{\frac{1}{R_1} + \frac{1}{R_2} + \frac{1}{R_3} + \frac{1}{R_4}} \qquad (3.15)$$

and is used with a lookup table to calculate c and d. Following Young (1989), the calculation is completed by evaluating

3.2 Analysis of Rolling Contact in Vacuum

$$\eta = \frac{K_D}{1.5}\sqrt{\left(\frac{1}{R_1}-\frac{1}{R_2}\right)^2 + \left(\frac{1}{R_3}-\frac{1}{R_4}\right)^2 + 2\left(\frac{1}{R_1}-\frac{1}{R_2}\right)\left(\frac{1}{R_3}-\frac{1}{R_4}\right)\cos 2\varphi},$$

(3.16)

where φ is the angle between planes of the elements 1 and 2. For rolling contact between the ball and rod, $\varphi = 90°$. The constants c and d are determined using a lookup table in Young (1989) based on the value of η. Next, we use Eqs. 3.12, 3.13, 3.14, 3.15, and 3.16 to calculate rod contact stress for the configuration shown in Fig. 3.2.

Example 3.2: Rod Contact Stress Calculation Consider the contact stress between a 12.7 mm diameter steel ball with elastic modulus $E_{ball} = 215$ GPa and Poisson's ratio of $\nu_{ball} = 0.29$ and a 9.53 mm diameter steel rod with elastic modulus $E_{rod} = 235$ GPa and Poisson's ratio of $\nu_{rod} = 0.29$, as shown in Fig. 3.2. The load applied to the cup, per ball, is given as $F_s = 23.83$ N. Compute the contact stress for this ball–rod combination.

Solution: The first step is to calculate the normal loads N_c and N_r based on an applied load F_s as shown in Fig. 3.2. Using Eq. 3.11 and a cup angle of $\beta = 25°$,

$$N_c = \frac{23.83 \text{ N}}{\tan(25°)} = 50.92 \text{ N},$$

(3.17)

and then $N_r = 2N_c$ or $N_r = 101.83$ N. The composite modulus is calculated using Eq. 3.3 as

$$E^* = \left(\frac{1-0.29^2}{215} + \frac{1-0.29^2}{235}\right)^{-1} = 122.59 \text{ GPa}.$$

(3.18)

Calculate the constant K_D using Eq. 3.15 as

$$K_D = \frac{1.5}{\frac{1}{6.35}+\frac{1}{6.35}+\frac{1}{4.76}+\frac{1}{(R_4 \to \infty)}} = 0.00286 \text{ m},$$

(3.19)

where the radius R_4 is very large to account for zero curvature in the axial direction of the rod. The constant η is calculated using Eq. 3.16 as

$$\eta = \frac{0.00286}{1.5}\sqrt{\left(\frac{1}{6.35}-\frac{1}{6.35}\right)^2 + \left(\frac{1}{4.77}-0\right)^2 + 2\left(\frac{1}{6.35}-\frac{1}{6.35}\right)\left(\frac{1}{4.77}-0\right)\cos \pi},$$

(3.20)

or $\eta = 0.40$. For clarity, a chart from Chap. 13 of Young (1989) for the values of α and β has been summarized in Table 3.1. From Table 3.1, $\alpha = 1.35$ and $\beta = 0.77$ for $\eta = 0.40$.

Table 3.1 Summarized chart from Table 33 of Young (1989)

Eta	0.00	0.10	0.20	0.30	0.40	0.50	0.60	0.70	0.75	–	–	0.98	0.99
Alpha	1.00	1.07	1.15	1.24	1.35	1.49	1.67	1.91	2.07	–	–	5.94	7.77
Beta	0.75	0.94	0.88	0.82	0.77	0.72	0.66	0.61	0.58	–	–	0.33	0.29

3.2 Analysis of Rolling Contact in Vacuum

The constants c and d are calculated using Eqs. 3.13 and 3.14, given as

$$c = 1.35 \sqrt[3]{\frac{101.83 \times 0.00286}{122.59 \times 10^9}} = 1.8 \times 10^{-4}, \qquad (3.21)$$

$$d = 0.77 \sqrt[3]{\frac{101.83 \times 0.00286}{122.59 \times 10^9}} = 1.03 \times 10^{-4}. \qquad (3.22)$$

Finally, the contact stress is calculated as Eq. 3.12 as shown:

$$\sigma_{rc} = \frac{1.5 \times 101.83}{1.8 \times 10^{-4} \times 1.03 \times 10^{-4} \, \pi} = 2.62 \text{ GPa}. \qquad (3.23)$$

Contact stress calculations are not exact, and it is therefore a good practice to do several calculations using different assumptions, but similar to the test condition. For example, the calculation in Example 3.2 was specific to contact stress between a sphere and rod. Consider also a calculation between two spheres of similar radii. Because the contact area of the ball and rod is less than the area between the contacting spheres, it is expected that contact stress between the ball and rod will be greater than between the spheres. This is illustrated next in Example 3.3.

Example 3.3: Contact Stress Calculation Between Two Steel Spheres For this example, we use the same material properties for the spheres as was used in Example 3.2 for the rod and sphere.

Solution: The composite modulus E^* will be the same as in Example 3.2, and the composite radius R is calculated using Eq. 3.1 as

$$R = \left(\frac{1}{6.35} + \frac{1}{4.76}\right)^{-1} = 2.72 \text{ mm}, \qquad (3.24)$$

which is the same as the calculation from Example 3.1, except that both spheres are made of steel. Assuming a circular contact area, the radius of the contact is given as

$$a_r = \left(\frac{3 \times 101.83 \times 0.00272}{4 \times 122.59 \times 10^9}\right)^{1/3} = 0.119 \text{ mm}, \qquad (3.25)$$

where for this calculation we used the contact force N_r and contact area instead of the normal pressure p_m as shown in Eq. 3.2. Now calculate the contact stress, or equally the contact pressure using Eq. 3.4, as

$$p_0 = \frac{3 \times 101.83}{2\pi \times 0.00019^2} = 1.35 \text{ GPa}, \qquad (3.26)$$

which is less than the contact stress of the ball and rod as expected. With this added calculation, we have more confidence in the calculation of Example 3.2.

As seen from Example 3.2, the contact stress calculation can be tedious. It is best to implement the calculation from Example 3.2 in a spreadsheet similar to Table 3.2 or some other code for systematic and repeatable calculation of contact stress. The calculation in Table 3.2 uses the same procedure outlined in Example 3.2.

3.2.2 Contact Load Versus Cycles

As pointed out earlier, the contact stress calculation is not exact in that many assumptions were made from the outset and it is sometimes more useful to report results in load versus cycles rather than stress versus cycles. It is certain that the applied normal load N_r in Fig. 3.2 is measureable and verifiable. Eventually, if all that is needed is a comparison of material or coating behavior, for example, steel from two different lots or coating from a "new" ion plating system, then reporting load versus cycles offers a straight forward comparison instead of involving assumptions about the contact stress in the comparison.

3.3 Rolling Contact Wear

The wear associated with rolling contact fatigue testing has been treated as a special case thus far in the chapter. Assumptions about contact area and material behavior are needed to complete the stress calculation, and the Hertz stress condition was assumed. But these assumptions were needed to satisfy a numerical explanation, specifically the contact stress calculation. It is equally valuable to have a general and phenomenological understanding of contact wear as well. For example, it is reasonable to assume that ball and rod temperature will increase with contact load. If we think of the test setup as a closed system, then the increase in contact load is one way to add energy to the system and that energy must be accounted for or dissipated in some way based on a thermodynamic approach. In this way, testing in rolling contact fatigue may be related to an analysis of two flat surfaces in friction except that in the case of flat surfaces, there is no subsurface maximum shear condition to initiate crack growth.

Let us think of a way to separate the wear and fatigue mechanisms associated with rolling contact fatigue. There is wear due to incipient sliding between surfaces, slightly to either side of the contact area between the ball and rod. Also, there is cyclic loading directly over the contact surface of the rod as the ball and rod rotate relative to each other. It would be advantageous to select a candidate material for the rod with input from basic material testing data such as the stress–strain and $S - N$ curves of the material. Then applying the appropriate correction factors based on temperature and geometry, we would have a pretty good idea of how the material will perform in rolling contact.

3.3 Rolling Contact Wear

Table 3.2 Example RCF spreadsheet calculation for contact stress

Item	Symbol	English units		Metric units	
Spring stiffness	k	75.60	lbs/inch	13.24	N/mm
Spring deformation	D	0.118	Inch	3.000	mm
Total test ball number	n	5		5	
Spring thrust load to single ball	Fs	5.36	lbs	23.83	N
Cup angle	β	25.083	Degree	25.083	Degree
Radial load between cup and ball	Nc	11.45	lbs	50.92	N
Radial load between rod and ball	Nr	22.89	lbs	101.83	N
Ball diameter	Db	0.5	Inch	12.70	mm
Rod diameter	Dr	0.375	Inch	9.53	mm
Ball contact diameter with cup	Dc	0.453	Inch	11.502	mm
Pitch diameter of balls and retainer	Dp	0.875	Inch	22.225	mm
Cup contact diameter with balls	Dcp	1.328	Inch	33.727	mm
Ball stress between cup and ball	σ1	199.23	ksi	1.402	GPa
Ball stress between rod and ball	σ2	388.11	ksi	2.659	GPa
Ball material elastic module	Eb	34,000	ksi	215	GPa
Rod material elastic module	Er	34,000	ksi	235	GPa
Cup material elastic module	Ec	29,500	ksi	235	GPa
Ball revolutions per retainer revolution	E	2.932	Revolution	2.932	Revolution
Rod revolutions per retainer revolution	F	4.910	Revolution	4.910	Revolution
Ball/rod stress cycles per retainer rev	H	3.910	Revolution	3.910	Revolution
Number of stress cycles between ball/rod per rod revolution	K	3.982	Revolution	3.982	Revolution
Rod rotational speed	cyc/s	130			
Total test time	hours	18.00			
Stopping threshold vibration	g	0.35			
Ball stress between rod and ball	Gpa	2.66			
Ball stress between cup and ball	Gpa	1.40			
Rod stress cycles	Nf	33540933			

3.3.1 Empirical Approach

The size and shape of the particles resulting from contact wear can give insight to the type and size of contact junctions between surfaces, Rabinowicz (1995). For rolling contact in vacuum, the sliding surfaces that experience the largest relative motion exist primarily at the edges of the wear track where the relative velocity of secondary contact of the elastically deformed ball contacts the rod. This wear situation is similar to incipient sliding in that the relative motion between the ball and rod is due to elastic deformation and not translational motion. The size of the contact junction is assumed to be the same as the particles generated from the contact. As part of a theoretical approximation of contact junction size, Rabinowicz (1995) presents the particle-size relation

$$d_A = 6.0 \times 10^4 \frac{U_{ab}}{H}, \qquad (3.27)$$

where H is the hardness of the softer contact material. The surface energy U_{ab} associated with the contact will vary depending on contact temperature, but it may be determined reliably if the test is operated in vacuum using solid film lubrication. Operating in high vacuum, one can be assured that surface oxide formation is minimal, and therefore a more accurate value of surface energy may be determined. The process for using Eq. 3.27 is to measure the diameter d_A of loose wear particles over a range of test temperatures, and then use that fit to predict surface energy U_{ab} associated with the formation of that particle. But the particle size is one part of a larger story.

Particle size analysis really gives more insight about coating adhesion and lubricity than the material response to fatigue wear. In fact, the surface energy U_{ab} in Eq. 3.27 is proportional to the elastic energy stored in the lattice structure of the contacting film. Analyses of this type assume an adhesive-wear condition and for testing in rolling contact fatigue, true particle size may be confounded since the particles from the contact are recirculated between the ball and rod contact areas throughout the test.

Electrical conductivity may be used for a detailed study of contact junctions in rolling contact fatigue. Some examples of sliding and rotating electrical contacts are relays and brushes on motors and slip rings. In particular, using a solid film of gold for lubrication instead of silver would enable more accurate resistance measurements between the balls and rod. For example, conductivity between the balls and rod will increase as the rotating elements run-in and then remain constant through the life of the test. Conductivity will decrease significantly when the gold lubricant is depleted near the end of the test. The decrease in conductivity is accompanied by an increase in mechanical vibration. Electrical conductivity between rolling elements is particularly important in rotating anode x-ray tubes. The rotating anode, shaft, and bearings are part of the electrical circuit needed to generate the x-rays. Copper also provides good electrical contact, but it is definitely not a lubricant for dry rolling contacts.

The friction generated during rolling contact fatigue testing with solid film lubrication has minimal effect on the fatigue-wear life of the rolling elements. In fact, the primary source of friction during the RCF is due to incipient sliding, and most often that friction will cause an adhesive failure within the film. Debris clutter generated from the contact friction will redeposit in the wear track between the ball and rod.

Let us now focus instead on the fatigue wear within the rolling elements due to cyclic contact in the wear track since it is this wear mechanism that leads to the maximum shear stress condition below the contact surface, as described in Example

3.1. Surface fatigue wear in rolling contact elements is more severe than ordinary fatigue wear of bulk materials. Surface fatigue wear involves the accumulation of many stress-loading and unloading cycles at the contact surface of the rolling elements. For comparison, fatigue in bulk materials will have no effect on life so long as the stress loading is 1/3 to 1/2 of the material yield strength, related to the so-called infinite limit. This is not the case for rolling contact elements and surface fatigue wear. Even for low contact loads, the repetitive application of stress causes subsurface cracks to form and ultimately leads to surface spall. Surface fatigue-wear data has enormous scatter as well, with variation in life of 200 to 1, Rabinowicz (1995). This is largely due to the confounding influences that temperature and internal material defects have on material properties of the contact elements. A life-load curve in the form of

$$t = \frac{\varGamma}{W^{\beta}}, \qquad (3.28)$$

where t is the time to failure and W is the contact load is useful to help reduce the effects of scatter in the data. The constant $\varGamma$ is specific to the size of the rolling elements and the characteristics of the test. The calculation in (3.28) is the basis for bearing-size selection found in mechanical engineering handbooks such as Shigley et al (2004) and Juvinall and Marshek (2011).

3.3.2 Thermodynamic Approach to Friction and Wear

Thus, far in this chapter, we have looked at wear and friction as a collection of empirical and theoretical concepts that can be verified and measured, but that do not share a unifying derivation. Let's change our thinking a little and consider friction to be a fundamental force in nature, as suggested in Nosonovsky and Mortazavi (2013). As such, friction would be inexplicably linked to the laws of thermodynamics and that linkage would enable a framework from which to select materials for testing in rolling contact fatigue. Self-organizing systems can at times bring symmetry and order to otherwise chaotic behavior. Nosonovsky and Mortazavi (2013) point to the Bernard cells problem as an example of how a system will self-organize to meet thermodynamic stability. In the case of heating and the Bernard cell problem, ordered convective flow of the heated fluid is the systems' way of dissipating heat energy.

Friction and wear can lead to material self-organization as well as a way to dissipate heat and stress. The new surfaces and even the internal spall cracks that form as a result of fatigue loading are thermodynamically stable. The concept of structural dissipative adjustment within the material has been investigated by Bershadsky and Kostetsky (1993). Recently, Amiri and Khonsari (2010) suggest that the formation of a tribo-film on wear contact surfaces is the material's response to external friction loads as well. These examples have been categorized as the material's synergetic response to friction and wear. In relation to rolling contact fatigue, the test may be considered "open" in that heat and mass are exchanged between the balls and rod through a mechanism known as third-body transfer. Then, a thermodynamic relation may be established for the friction and wear situation as

$$dU = -PdV + TdS + \psi dN, \quad (3.29)$$

where P, T, and ψ are the pressure, temperature, and chemical potential of the open system. The amount of material at any given state N is the number of particles that will be influenced by the analysis conditions, as multiplied by the chemical potential ψ. The change in energy dU of the system due to friction and wear is analogous to mechanical work and energy to the system as $dU = -Fdx$. Entropy accounts for the irreversibility and dissipation of energy in the rolling contact system. As such, it may be used to characterize material behavior during testing. All unknown variables and effects to the system are accounted for in the entropy term S. Using Eq. 3.29, one can apply thermodynamic analogies to the rolling contact fatigue-wear situation. For example, a thermodynamic stability criterion based on the second variation of the change in entropy may be stated as

$$\delta^2 \dot{S} > 0, \quad (3.30)$$

which is based on the minimum entropy production principle, as applied by Nosonovsky and Bhushan (2009). The differential operator δ is used so that Eq. 3.30 may be applied to any general case. If the entropy production rate is constant and a minimum, then the variation of entropy production will be zero or $\delta \dot{S} = 0$. Now define the entropy rate in terms of mechanical interactions that are known to exist during rolling contact. Let the entropy produced in the rolling contact system be a function of friction force, load, and temperature or

$$\dot{S} = \frac{\mu W \dot{x}}{T}, \quad (3.31)$$

where the friction force μW is applied over a velocity between surfaces as $\dot{x}$ at a given temperature T. If only one of these is varied, then either $(\delta W)^2 > 0$, $(\delta \dot{x})^2 > 0$, or $(\delta T)^2 > 0$ will always be true and the stability will be as

$$\frac{\partial^2 S}{\partial W^2}(\delta W)^2 > 0, \text{ or } \frac{\partial^2 S}{\partial \dot{x}^2}(\delta \dot{x})^2 > 0, \text{ or } \frac{\partial^2 S}{\partial T^2}(\delta T)^2 > 0. \quad (3.32)$$

The system will remain stable. But if friction force μW and velocity $\dot{x}$ are varied simultaneously, and we know that they are not independent of each other, then using Eq. 3.31 the stability calculation becomes

$$\delta^2\left(\frac{2\mu W\dot{x}}{T}\right) = \frac{2\mu}{T}\delta W\delta\dot{x} = 2\mu\frac{dW}{d\dot{x}}(\delta\dot{x})^2, \tag{3.33}$$

where stability is now dependent on the sign of dW and $d\dot{x}$ and system stability is no longer guaranteed. If fact, if the rate of change of either the friction force or velocity decreases as a result of the variation of the other, the system will be become unstable. To put this in context of rolling contact fatigue in high vacuum, this instability will force a change or new equilibrium within the contact area, leading to the formation of a subsurface spall or a new surface in the contact area. More thermodynamic stability analyses related to friction and wear of sliding surfaces are presented in Nosonovsky and Mortazavi (2013). These examples may be validated using the rolling contact fatigue platform in high vacuum that is presented in Chap. 4. Operating in high vacuum reduces confounding results due to surface contamination and the influence of surface oxides. Table 5.1 in Nosonovsky and Mortazavi (2013) lists other entropic definitions and equations that enable a connection between wear and friction phenomenon.

References

Amiri M, Khonsari M. On the thermodynamics of friction and wear – a review. Entropy. 2010;12:1021–49.

Bershadsky L. B. I. Kostetski and the general concepts in tribology. Trenie I Iznos. 1993;14:6–18.

Beswick J. STP1548. West Conshohocken: ASTM; 2012.

Bhushan B. Principles and applications of tribology. New York: Wiley; 1999.

Holmberg K, Mathews A. Coatings tribology: properties, techniques and applications in surface engineering. Danvers: Elsevier Science; 2009.

Juvinall R, Marshek K. Fundamentals of machine component design. 5th ed. New York: Wiley; 2011.

Mattox D. Handbook of physical vapor deposition processing. Westwood: Noyes; 1998.

Mattox D, Kominiak GJ. Structure modification by ion bombardment during deposition. J Vac Sci Technol. 1972;9(1):528–32.

NASA SP-5059(01). Solid lubricants: a survey, technology utilization office NASA, NASA; 1972.

NASA-2011-216098. Rolling-element fatigue testing and data analysis-a tutorial. Technical report, Cleveland: NASA Glenn Research; 2011.

Nosonovsky M, Bhushan B. Thermodynamics of surface degradation, self-organization and self-healing for biometric surfaces. Philos Trans R Soc A Math Phys Eng Sci. 2009;367:1607–27.

Nosonovsky M, Mortazavi V. Friction-induced vibration and self-organization: mechanics and non-equilibrium thermodynamics of sliding contact. Boca Raton: CRC Press; 2013.

Rabinowicz E. Friction and wear of materials. New York: Wiley; 1995.

Shigley JE, Mischke CR, Budynas RG. Mechanical engineering design. 7th ed. New York: McGraw-Hill; 2004.

Stanchowiak G, Batchelor A. Engineering tribology. Burlington: Elsevier Butterworth-Heinemann; 2005.

Van Overschee P, De Moor B. N4SID: subspace algorithms for the identification of combined deterministic-stochastic systems. Automatica. 1994;30:75–93.

Young W. Roark's formulas for stress & strain. 6th ed. Vols. Chap. 6,7,10. New York: McGraw-Hill; 1989.

Chapter 4
Rolling Contact Fatigue in High Vacuum

Abbreviations

L_β	Life at which 63.2 % of samples failed
L_{50}	Life at which 10 % of sample failed
L_{10}	Life at which 50 % of sample failed
AES	Auger electron spectroscopy
CVFC	Control volume fraction coverage
MLE	Maximum likelihood estimation
RCF	Rolling contact fatigue
SEM	Scanning electron microscope

Thin-film coatings on the order of nanometers in thickness are particularly susceptible to atmospheric contamination which may negatively impact both their composition and performance. Testing in a high-vacuum environment is desirable to get an accurate assessment of performance and may also be used to quantify the effects of process behavior on a coating performance. In later chapters, we will explore how deposition process variables such as pressure and voltage influence the performance of a solid silver lubricant applied to ball bearings that operate in high vacuum. Effects of these process variables on the lubricating film composition will be discussed as well.

The goal of this chapter is to design a test setup that will replicate, as closely as possible, the loading and environment condition in which the film will be used in service. For example, if a solid film lubricant is to be used for a bearing application inside a high-voltage and high-vacuum device, such as a rotating anode x-ray tube, rolling contact fatigue testing in vacuum is among the best test method to check in-service performance of the bearing components. Using the RCF test method, one may evaluate the rolling contact fatigue life of a candidate lubricant coating as well as quantify the effects of process variable aberration on coating life. A stable, repeatable, and validated platform from which to make an assessment is needed. In this chapter, we present a test rig specifically designed to test solid film lubricants in rolling contact fatigue under high-speed and high-vacuum conditions.

4.1 Introduction

Life testing of thin solid lubricating films for high-cycle fatigue applications is best achieved using the rolling contact fatigue (RCF) test method. Whether testing solid lubricant films or wear resistant hard coatings, the RCF method allows rapid accumulation of load cycles in less time than it would take to test the coating system in its in-service application. Specifically, the contact loading may be increased to shorten test length while still providing valuable insight to how the film lubricant will perform. For ball bearing systems that operate at high speed and in high vacuum, the RCF test method closely replicates dynamic loads due to high-speed rotation.

Film lubrication testing using in-service bearings parts reduces development costs. It is less expensive to test specific aspects of a rotating ball bearing system rather than the entire system at once. If the coated bearing elements are part of a larger system, as in a high-voltage and high-vacuum device such as a rotating anode x-ray tube, it can be risky to install new lubrication coatings into the larger system with little or no information about performance. What if a component of the system, other than the bearing, fails before the coating under test? It is best to test a new coating and ball bearing assembly in isolation from the rest of the system to enable non-confounding development of the coating.

Analysis tools such as Auger electron spectroscopy (AES) and scanning electron microscopy (SEM) may be used to assess coating composition and thickness over time. It is extremely valuable to suspend a test for analysis prior to failure. Results from these analysis methods may be used to validate predictive models with focus on simulating the coating process to develop adequate process control schemes. The ideal scenario is to stop and autopsy a few tests just before failure, in order to evaluate the film lubricant L_β life. The test scheme should consist of a least 15 tests, with at least 1/3 of them run to failure. Knowing the L_β, L_{50}, and L_{10} life of the test elements, one may use statistical tools to correlate with in-service bearing systems.

The RCF testing method was first developed using oil-based liquid lubrication. However, if the coating under test needs to operate in high vacuum, a thin solid film of an inert lubricant will work as well. Testing using a thin solid lubricant in high vacuum allows for very high-speed testing and therefore rapid accumulation of stress cycles compared to testing the same coating with oil-based lubricants. Oil-based lubricants can foam which may introduce hydrodynamic instabilities at rotational speeds approaching 1,000 rpm, which will confound the test results.

4.2 Rolling Contact Fatigue Test Platform

High-voltage and high-vacuum devices such as rotating anode x-ray tubes operate in the range of 7,000–10,000 rpm, and therefore thin solid lubricants must be used in the ball bearing system under test.

Application of a solid film prior to testing allows a quantitative measure of the amount of lubrication applied to the system. For comparison, testing with oil allows for a seemingly endless supply of lubrication, as the oil is allowed to flow repeatedly over the rolling elements. Indeed, testing with liquid oil-based lubricants increases test longevity, but it also increases the time it will take for the rolling elements to fail in fatigue. Testing with oil lubrication is not considered in the present chapter because it does not represent the operating environment of the thin-films system under consideration, namely, solid lubricants for operation in high vacuum and high rotational speed conditions. For more discussion on RCF testing in oil, see NASA-2011-216098 (2011).

Load adjustment in the RCF test fixture allows control of test length. For example, a lightly loaded test using 12.7 mm diameter steel balls with silver film lubrication may go up to 25 h rotating at 7,800 rpm. In comparison, a heavily loaded test of the same ball size, lubrication, and rotational speed may fail within 5 h. As with all sub-element testing, a transfer function needs to be established between test length and the longevity of the in-service bearing system. Historically, RCF test-life data is highly scattered, and therefore statistical analysis tools should be used to draw conclusions concerning coating life and how it correlates with in-service bearings.

The RCF test method in high vacuum enables continuous testing in a reaction-free environment. It may be used to study changes in microstructure and chemistry as a result of repeated contact stress. Chemical changes in the film due to heat from friction and cyclic fatigue loading may be evaluated without concern of carbon atoms commonly found in oil-based lubricants. Surface modification of the rolling elements due to contact loads alone may be studied using the RCF method in vacuum with a solid silver or similar film as lubrication.

Scratch testing and hardness testing methods do not test the film in high-cycle fatigue loading. The effects of dynamic loading and vibration due to the rotational aspect of the RCF test method also enables one to account for conditions related to rotating machinery, such as ball-pass frequency. One may get positive results from a scratch and hardness test in a laboratory environment, only to find that the coating flakes off during in-service operation of the bearing system.

4.2 Rolling Contact Fatigue Test Platform

All ball bearings and rolling elements contain surface defects described as surface asperities. The peak height and distance between asperities is a key attribute of a contacting surface. The asperity peaks may be quantified using surface analysis tools such as a surface profilometer or a scanning electron microscope (SEM). The surface is mapped for relative height of peaks and valleys and then assigned a Ra number related to the height of the peak, or asperity, and the width between peaks. For more information concerning defects and classification, see Bhushan (1999) and Totten and Liang (2004).

Surface lubrication, whether solid or liquid, may be divided into three categories: (i) full film, (ii) boundary layer, and (iii) mixed film and boundary layer. Full-film lubrication is the condition such that the film thickness is greater than the height of the surface asperities. Load is transferred through the film, and the asperities on either surface do not contact each other. Full-film lubrication is the ideal operating condition for any rolling-element bearing system. Boundary layer lubrication describes the condition such that load is transferred by asperity-to-asperity contact. The lubricant film is present in the valleys between asperities, but the load is transferred by peak-to-peak contact. The mixed film and boundary layer situation is a combination of full film and boundary layer in which there is load transfer by both the film and asperity contact. One advantage for using solid lubricants to study bearing-element contact behavior is that the transition between these three categories is easily detected using accelerometers and vibration detection equipment. Later in this chapter and in Chap. 6, we present test data in which vibration measurement was used to observe the transition through these three film categories.

Rolling contact friction of a ball on a flat surface will generate an elliptical contact region due to elastic deformation of the ball. The coefficient of friction of the contact is defined as

$$\mu_r = \frac{3\alpha a}{16R}, \qquad (4.1)$$

where a is the half-width of the contact zone, R the radius of the ball, and α the loss factor due to elastic hysteresis of the ball material. Concerning RCF loading conditions, subsurface cracking beneath the contact area is the source of spall and eventual flaking of the contact surface. The initiation of the subsurface crack is caused by the high compressive stress at the contact area that results in a maximum shear stress condition immediately below the contacting surface. One method to reduce subsurface crack initiation is to apply a lubrication layer to reduce friction and thereby reduce the subsurface shear stress.

Surface condition can influence lubrication performance. For example, the surfaces of rolling elements may contain physisorbed layers of contaminant atoms and molecules that are held in place by weak van der Waals bonds. As the surface is loaded and it heats up, these layers can evolve into the lubricant and upset the structure of the lubricating film. When testing in high-vacuum conditions, the liberated physisorbed atoms and molecules can increase chamber pressure and are detectable using RGA equipment.

The energy to initiate physisorption outgassing is about 2 kCal/mol. The molecules and atoms of the physisorbed layer do not share electrons with the surface atoms of the rolling elements, and for that reason, they are easily knocked-off from the rolling surface. In contrast, chemisorbed layers of atoms and molecules do share

4.2 Rolling Contact Fatigue Test Platform

electrons in their outer shells with the rolling-element surface atoms and may have an associated bonding energy of 10–100 kCal/mol. The chemisorbed layer has a more significant influence on surface contact performance due to its higher bonding strength compared with physisorbed molecules. The presence of carbon as well as numerous types of oxides will influence test results as well. It is very important to know what is happening inside the chamber during testing, lest the presence of an oxide positively or negatively influences test life. While it is impossible to assemble a test free of carbon and oxide contamination, it is possible to reduce their presence by heating the test chamber and components to speed up the outgassing process before starting the test. A through outgassing procedure immediately before the test will almost completely remove their influence when operating inside a vacuum chamber.

Friction in RCF testing arises due to elastic deformation of the contacting surfaces. Deforming surfaces dissipate energy and generate heat internally, and sliding at the outer edges of the contact area will contribute to heat generation as well. Hertz contact analysis is a common framework on which RCF test results may be evaluated. Hertz contact analysis assumes small deformations in the contacting areas that are within the elastic limit of the materials. The surfaces are assumed continuous and nonconformal, that is, they do not fit together like pieces of a puzzle. Finally, the contact area is much smaller than the radius of the contacting elements, so that the resulting contact approaches a flat-contact situation within the contact zone (Bhushan (1999)).

If a Hertz contact analysis is used to calculate RCF stress, the load associated with subsurface yield of the contacting elements may be related to the applied normal load. For example, the contact area a associated with the contact of two spheres of radius R_1 and R_2 with applied force W is

$$a = \left(\frac{3WR}{4E^*}\right)^{\frac{1}{3}} \tag{4.2}$$

where the effective radius of contact R is calculated as

$$\frac{1}{R} = \frac{1}{R_1} + \frac{1}{R_2}. \tag{4.3}$$

The composite modulus E^* is defined as

$$\frac{1}{E^*} = \frac{1-\nu_1^2}{E_1} + \frac{1-\nu_2^2}{E_2}, \tag{4.4}$$

using the Poisson ratios ν_1 and ν_2 of the contacting materials. The pressure associated with the contact may be defined as

$$p = p_0 \left\{ 1 - \left(\frac{r}{a}\right)^2 \right\}^{\frac{1}{2}}, \tag{4.5}$$

where the maximum contact pressure p_0 may be defined as a function of the normal load W and the radius of the contact area a. The maximum contact pressure is given as

$$p_0 = \frac{3W}{2\pi a^2}. \tag{4.6}$$

Equations 4.2, 4.3, 4.4, 4.5, and 4.6 may be used to set the upper limit for a RCF test.

Fatigue failure beneath the surface of a rolling contact is thought to proceed as follows: (i) work hardening, (ii) elastic response, and (iii) material softening and yielding (Sadeghi et al. (2009)). If a solid lubricant film such as silver or gold, or a similar lubricant, is present on the contact surface, then the work hardening step involves blunting of asperities along with material transfer of the solid film back and forth between surfaces. The coating material transfer process will be explained later in the chapter using a third-body model transfer mechanism. So long as sufficient film is present to enable full-film and even mixed-boundary lubrication, the contacting surfaces respond elastically to the contact load and the RCF test will continue without spall. However, as the lubricating film is depleted, the material in the contact area will begin to yield plastically due to increased friction and shear stress. The volume of plastically yielded material increases as the lubrication or film is depleted. Plastic yielding begins the onset of subsurface crack initiation and continues with repeated load cycles. Eventually, coating spall will occur at the contact surface which further increases shear loading beneath the surface of the ball and furthers the onset of crack initiation below the surface. Subsurface cracking and spalling may also result from internal material defects. The presence material defects in the lattice structure of the contacting materials of both the rolling elements and the lubricating film is a large contributor to the wide scatter in RCF test results.

The RCF method in this chapter uses ball bearings and a rotating rod to load thin-film coatings in high-cycle fatigue. Coatings on either the balls or the rotating rod may be tested as shown in Fig. 4.1. The RCF test method was first conceived in 1958 by W. J. Anderson at NASA Lewis for gear and bearing research. Originally named the "NASA 5-ball fatigue tester," Anderson constructed the tester in 1959, and today it is simply known as the ball–rod–cup RCF tester. The ASTM publication STP771 recommends ½" diameter balls made from M50 tool steel be used for consistent testing of candidate rod materials. The ball–rod–cup test configuration has matured such that the balls and cups may be purchased from leading bearing component manufacturers. The test rods are typically made by the investigator to study material response to rolling-element-type loading. However, the balls and

4.2 Rolling Contact Fatigue Test Platform

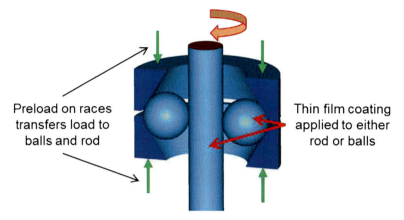

Fig. 4.1 Cross section of ball–rod–race loading and thin-film locations

cups may be any custom size and material type as well, with any type of lubricating coating needed to replicate a specific test need.

There is a wealth of test data concerning the RCF test method in oil. For example, the test rig configuration called out in Hoo (1981) has been used to quantify ball–rod combinations and surface finishes and their effects on material fatigue life. In Hoo (1988), the RCF method was used to quantify the effects of material processing on tool steel. In Hoo (1988), bearing steels are tested in rolling contact fatigue to better understand how process aberration and lattice structure defects influence the fatigue life of steel. Beswick (2012) illustrates several configurations used for RCF testing. All of these procedures callout oil film lubrication, and, in particular, Beswick (2012) recommends a specific drip rate for the oil along with test-temperature recommendations. The 3-ball and 5-ball testers developed by J.C. Hoo and W.J. Anderson have been very popular through the 1980s. These test rigs have been used successfully to qualify steel manufacturing processes. Adaptation of the RCF test method for high rotational speed in high-vacuum conditions was first published in Danyluk and Dhingra (2012a). The introduction of this platform for high-cycle fatigue testing in high vacuum enables study of deposition methods on coating life in an inert atmosphere.

4.2.1 Test Configuration in Oil

A bronze cage is used in Hoo (1981) to separate three balls of 12.7 mm diameter while rotating with the rod. Oil provides hydrodynamic lubrication between the balls, rod, and cup for the duration of the test. From ball and rod diameter comparison, the RCF test configuration in Hoo (1981) is designed to apply at least twice the contact stress to the rod surface as compared with the surfaces between the balls and cups. The contact stress increases for decreasing contact radius, based on a Hertz contact analysis, and there are requirements for the oil used in the test. The general idea of the ball–rod–cup RCF test is that the rod will be made from the candidate material for fatigue testing and it will fail before either the balls or races.

4.2.2 Test Configuration in Vacuum

An RCF test platform that can uniquely test coating adhesion and film quality under RCF conditions in high vacuum and at high rotational speeds is presented in this section. The test platform is a modification of the work of J.C. Hoo (1982) to allow testing in high vacuum and at 7,800 rpm rotational speed. The vacuum chamber and pumps may be purchased from leading vacuum hardware suppliers. The chamber assembly is illustrated in Fig. 4.2, along with information about the top-section attachments. A large view port was installed to enable observation during testing as well as optical temperature measurement of the test fixture. The smaller ports, 2.75 and 1.33 in., are used for passing vacuum measurement test probes and for installation of a cathode for electron-beam heating of the rod during testing. Figure 4.3 presents an assembled chamber with supporting pumps and valves underneath. The test elements, balls, rods, and cups, are shown in Fig. 4.4. The RCF fixture and thermocouple are positioned inside the vacuum chamber as shown in Fig. 4.5. The RCF fixture rests on the floor of the chamber and is not rigidly attached. A fixture wire is used to prevent rotation of the fixture as friction increases near the end of the test.

The bronze carrier cage specified in Hoo (1981) and Beswick (2012) was not used for testing in vacuum. The contact of the carrier cage with the balls under test allowed bronze material transfer to the coating under test and contaminated the coating. Specifically, the bronze material mixed with the silver film on the balls and this condition no longer represented the desired test. From henceforth, the bronze carrier cage was not used. Instead, five balls of diameter 12.7 mm as described in configuration 1 of Table 4.1 were successful. A second attempt with a ball cage was made using a 3-ball carrier cage fabricated from 304 stainless steel, chosen for its inert properties. However, the stainless steel carrier cage produced too much vibration at 130 Hz (7,800 rpm) rod rotation such that the vibration threshold for spall detection was exceeded and the test needed to be stopped. For comparison, the nominal rotational speed of Hoo (1981) and Beswick (2012) is 3,600 rpm, compared with testing in vacuum using silver film lubrication rotating at 7,800 rpm.

A servo motor was used to rotate the rod as well as track total test time before the onset of failure. The drive motor is mounted outside the chamber, and motor torque is delivered to the rod inside the vacuum chamber using a ferrofluidic rotary feedthrough device. High vacuum is applied using an Agilent^{TM} V-81 turbo pumping system as shown in Fig. 4.3. A vibration monitor is used to continuously collect accelerometer data. The vibration monitor is connected in series with the drive-enable circuit of the servo drive. When the vibration threshold is exceeded, the monitor will open the circuit and disable the servo drive, effectively stopping the test.

4.2 Rolling Contact Fatigue Test Platform

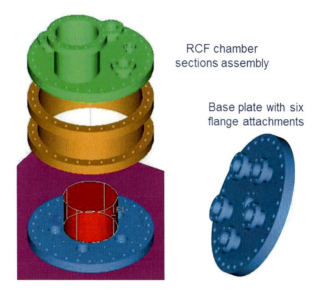

Fig. 4.2 RCF test chamber section assembly order, with custom base plate for vacuum component attachments and diagnostics

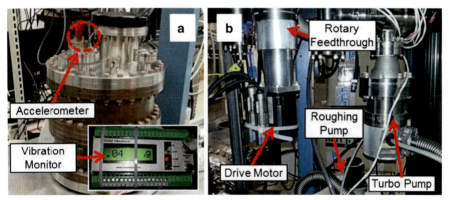

Fig. 4.3 (a) RCF chamber, accelerometer, and vibration monitor, (b) support hardware underneath: turbo and roughing pump system, motor drive, and ferrofluidic rotary feedthrough

4.2.3 Rolling Elements in Vacuum

Test longevity and the coating's ability to endure RCF cycling are dependent on ball and coating properties as well as the test configuration. For example, the thin film deposited on 12.7 mm diameter balls is likely to spall if the applied contact stress approaches the yield stress of the ball material. When planning a RCF test, it is best to measure the surface hardness of the components rather than to rely on tabulated published data for the bulk material. The ball, race, and rod sizes and curvatures influence the measured hardness of each test element.

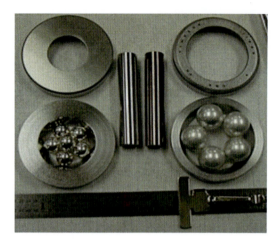

Fig. 4.4 RCF test elements: Rex 20 and Si_3N_4 rods with M50 and Rex 20 balls races (Reproduced with permission from Wear, Volumes 274–275, 27 January 2012, Pages 368–376, Elsevier)

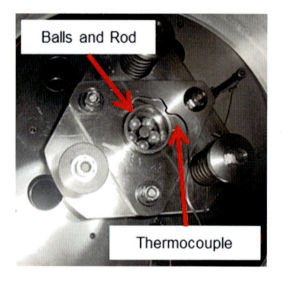

Fig. 4.5 RCF test fixture inside vacuum chamber with thermocouple placement

Two ball sizes of two different materials were used to validate and commission the RCF platform for high-vacuum testing. The rod and cup materials were varied as well to enhance the validation of the new RCF platform. Testing with 12.7 mm diameter balls is an established practice for the ball–rod–cup test rig and also enabled a sanity check for test cycle length compared with testing in oil. For the RCF testing in this chapter, the rod diameter is always fixed at 9.53 and 50.8 mm in length. The race material was varied between Rex 20 steel and ANSI M50 steel. The calculated Hertz contact stress on the balls for all tests was between 2.1 and 4.2 GPa. This stress range is consistent with previous testing using ANSI M50 steel.

In addition to testing with 12.7 mm balls, a test procedure using six 7.94 mm diameter balls was developed as described in configuration 2 of Table 4.1. Testing

4.2 Rolling Contact Fatigue Test Platform

Table 4.1 Ball, rod, and race configurations for RCF high-vacuum testing

Config.	Ball material	Rod material	Race material	Lubrication	Ball size	Number of balls	Number of tests
1.	M50 steel	Rex 20	M50 steel	Silver	12.7	5	33
2.	ANSI T5	Rex 20	Rex 20	Silver	7.94	6	27
3.	M50 steel	Si$_3$N$_4$	M50 steel	Silver	12.7	5	17
4.	M50 steel	Rex 20	M50 steel	Silver	12.7	3 balls with 304SS cage	2

Reproduced with permission from Wear, Volumes 274–275, 27 January 2012, Pages 368–376, Elsevier

Table 4.2 Hardness and material property data for ball, rod, and race components

Material property	9.53 mm diameter Rex 20 rod	9.53 mm diameter Si$_3$N$_4$ rod	7.94 mm steel ANSI T5 ball	12.7 mm steel M50 ball	Rex 20 race	M50 race
HRC (measured)	62.9	74.6	61.8	62.1	66.2	42.4
Elastic modulus GPa	235	310	214	203	235	203
Poisson ratio	0.29	0.25	0.29	0.29	0.29	0.29

Reproduced with permission from Wear, Volumes 274–275, 27 January 2012, Pages 368–376, Elsevier

with 7.94 mm balls was used to help set a baseline for test length for the RCF test rig. It is possible to purchase pre-coated, 7.94 mm diameter balls from an established bearing and thin-film coating supplier. The purchased pre-coated balls have an established history concerning operation in high-vacuum inside rotating anode x-ray tubes. Test rods made from Si$_3$N$_4$ were tested as well as shown in configuration 3 of Table 4.1. Table 4.2 presents hardness and material property data for all test elements, and Table 4.3 contains measured surface roughness data. The hardness and roughness measurements were repeated three times per sample on five samples, and the average was taken based on 15 measurements of each test element. Material property values may be highly scattered, and yet it is not practical to measure properties on every test element ever tested. Rather, a sample measurement size of 15 was chosen to give statistical confidence of the measured data in Table 4.2.

Table 4.3 Surface roughness Ra data in microns for ball, rod, and race components

12.7 mm M50 ball	M50 race	7.94 mm T5 Ball	Rex20 race	9.53 mm Si_3N_4 rod	9.53 mm Rex20 rod
0.32	0.37	0.05	0.12	0.04	0.13

4.3 Rolling Contact Fatigue Vacuum Test

Process history and test-element preparation will have a significant influence on RCF test results. Thin solid films on the order of 500 nm thick or less are at risk for exposure to air or contact with volatile compounds during handling. For example, a silver film that has come into contact with low-melting metals such as tin, indium, or lead will itself have a lower melting temperature due to contamination and will therefore give misguided results. The contaminant metal compound from contact with the bronze carrier cage in Sect. 4.2.2 significantly increased test length, which was not the intent of the test. Silver film coatings on the balls that are exposed to air will collect water vapor molecules, and that water vapor will react with surfaces of the test elements inside the vacuum chamber.

Component cleanliness and honest adherence to established vacuum practice as outlined in handbooks such as Mattox (1998) should be implemented in order to have confidence in the test result. For all RCF tests carried out in this monograph, all test elements were cleaned in an ultrasonic bath of methylene chloride for 20 min to remove surface oils and particulate contamination. After cleaning, all parts should be stored in heated, dry nitrogen storage or if possible under high-vacuum conditions to reduce surface–gas accumulation to the test elements. Alternatively, the cleaned parts may be stored in an appropriately lined vacuum bag under vacuum. A note of caution is needed. Not all vacuum bags are the same. One has to make sure that the lining of the bag does not collect and hold moisture while exposed to air. Experience has shown that moisture trapped inside a bag will react with metal parts even though the inside of the bag is under vacuum.

4.3.1 Coated Rolling Elements in Vacuum

All coated balls used in this chapter were coated with silver in high vacuum, about 10^{-6} Torr. Silver film was evaporated on to the ball surfaces without the application of ion implantation. The goal of this chapter is to validate the RCF test platform for use in high vacuum. Influence of ion implantation is another level of complexity that will be addressed in Chaps. 6 and 7. Thin-film silver lubrication was chosen for its inertness and for its load and temperature capabilities while operating in vacuum and at high temperature.

There is a wealth of experience in the thin-film community concerning application of silver films. For example, silver is used on the threads of ultra-clean stainless

4.3 Rolling Contact Fatigue Vacuum Test

steel hardware such as nuts and bolts to prevent galling. Copper gaskets may be coated with silver to prevent copper oxide formation on the exposed surfaces of the gasket inside the chamber. For example, if the chamber is to be opened to air on a regular basis, then the exposed copper surface of the gasket should be coated with silver to prevent formation of copper oxide.

Prior to application of the film, the balls were outgassed inside the coating system chamber with vacuum in the range of 5×10^{-7} Torr. The balls were kept at about room temperature and therefore the outgassing was continued for 24 h prior to silver coating. Even though the balls were cleaned in methylene chloride and stored in nitrogen, a thin layer of physisorbed water molecules will remain on the surface. Outgassing in high vacuum allows most of the water to evaporate away from the surface before coating. Alternately, heat may be used to speed up the outgassing process.

Immediately before the coating is applied, the balls were further cleaned using an argon-ion scrubbing process within coating chamber. PVD coating experts such as Mattox (1998) recommend argon-ion processing immediately before deposition. The ion-scrubbing process is similar to an atomic sandblaster in which ionized argon atoms are used to knock off any remaining water vapor molecules from the ball surface. A note of caution though, too much ion scrubbing can lead to elemental mixing and contamination implantation. Approximately 200 nm of silver was evaporated on to the ball surface. During the coating process, the balls were agitated in a carousel positioned above the crucibles containing silver to insure uniform deposition thickness.

4.3.2 Test Assembly in Vacuum Chamber

Pre-coated ANSI T5 balls, 7.49 mm in diameter, were tested and compared with other coated balls. Adding pre-coated balls to the test validation process will help to establish a baseline for operation of the RCF test rig. The pre-coated balls were purchased from an established industrial ball bearing supplier and were kept in a vacuum storage bag until testing. However, each time the balls and other test elements are exposed to air, the surfaces need to be outgassed prior to starting the test. For example, after the hardware is loaded into the RCF chamber, the system is outgassed at room temperature for 12 h at less than 5.0×10^{-7} Torr vacuum before starting the test. All tests were carried out in the chamber shown in Fig. 4.5 and in the vacuum range of 10^{-5} to 10^{-7} Torr.

4.3.3 RCF Test Failure Criterion

Experience with the test rig in Fig. 4.3 and in particular with the location of the accelerometer on the top of the chamber confirms that the onset of a spall on at least one ball is detectable over a vibration range of 0.22–0.35 g. As part of the validation of the RCF test rig, the accelerometer location was the same for all tests. For the first 15 min of the test, a run-in period is observed in which the vibration levels may reach 0.30 g. This corresponds to the shakedown of contacting elements in which asperities are reshaped. It is also the time in which the solid film is passed between contacting surfaces, between the balls, the rod, and the races. After about 15 min, the vibration amplitude decreases to a steady-state range of 0.06–0.15 g for the majority of the test. As the test progresses, the solid film lubricant becomes depleted and the vibration steadily increases, corresponding to a mixed film and boundary contact condition.

The temperature of the test fixture increases with increasing accelerometer measurements and tended to increase over the length of the test. This is to be expected since as the test progresses and the silver is depleted from the contact surface, there is a corresponding increase in friction and vibration response leading to an increase in temperature. The temperature is measured using the thermocouple located as shown in Fig. 4.5. The thermocouple passes through the top section of the RCF fixture and is in contact with the top cup. The peak temperature measured at the cup during failure is repeatable at about 120 °C. However, for most of the duration of the test, the cup temperature ranges from 80 °C to 100 °C as measured at the location shown in Fig. 4.5. The estimated ball temperature at failure corresponding to the accelerometer measurement of 0.35 g is approximately 160 °C. This measurement is done using an optical pyrometer focused on the 12.7 mm M50 balls.

4.3.4 Analysis Tools

Statistical tools are needed to extract useful information from the RCF test results. For example, a Weibull analysis found in high-cycle fatigue software such as Weibull^{++} from ReliaSoft^{TM} may be employed to compare test results independent of coating and test element characteristics. An inverse power law, in combination with the Weibull distribution, may be used to check the validity of the test results. Specifically, the Weibull parameters, shape and scale, can give insight about test results. Additionally, Weibull^{++} is used to establish cycles versus failure and reliability curves based on at least two test group loadings. For example, to estimate the life-cycle curve for any RCF data, testing is needed for at least two contact stress levels. A Weibull distribution is used to characterize the test data at each stress level, and a line may be drawn between the means of the two distributions. The curve connecting the groups is the load versus cycles approximation curve.

Accelerated stress testing of mechanical component allows rapid accumulation of test cycles in a shorter period of time compared to testing at in-service stress levels. Consider the comparison of walking or running 5 km; the amount of work performed is the same, but the energy expended is not. More energy is expended in running than walking the 5 km. For testing mechanical components, the inverse

4.3 Rolling Contact Fatigue Vacuum Test

power law model may be applied to accelerated stress testing. The inverse power law is

$$L(V) = \frac{1}{KV^n}, \qquad (4.7)$$

where L is the life value of the test or time to failure. The parameters V, n, and K are the stress level and model parameters of the test. These parameters are fit to the data using the power law model equation. The power law may be combined with the Weibull distribution to form the inverse-power law-Weibull (IPL-Weibull) distribution, given as

$$f(t, V) = \frac{\beta K V^n}{(KV^n t)^{\beta-1}} e^{-(KV^n t)\beta}. \qquad (4.8)$$

Careful observation of the IPL-Weibull distribution reveals that it is the Weibull distribution using the substation $\eta = L(V)$. This is a three-parameter model, that is, the test data is used to determine values of K, β, and n. Weibull++ will solve for these parameters automatically using a parameter-estimation algorithm based on the maximum likelihood estimation method (MLE). In general, it is a least squares fit of the test data to three partial derivative equations of MLE function, similar to the concept of moment-generating function used in statistical analyses. A typical data set from RCF testing may have the form shown in Table 4.4. Some of the analysis tools within Weibull++ are presented next in Example 4.1.

Example 4.1: Data Analysis Using Weibull++ The data set in Table 4.4 is loaded into a Weibull++ project as shown in Fig. 4.6. First, the model type is chosen to fit the data. The inverse power law is a good model for cyclic loading of mechanical elements such as bearings and gears. Next, a distribution function is selected. At this point, little is known about the data, and the Weibull distribution is commonly used as a starting point since it offers the most flexibility for all data distribution types. With the model and the distribution type selected, the model parameters are calculated as shown in Fig. 4.7. The software automatically calculates the parameters discussed in Eqs. 4.7 and 4.8. Weibull++ is a useful tool for analysis of RCF data sets as well as for use with other fatigue-type loading data processing.

The data set given in Table 4.4 was collected from 33 RCF tests over two stress levels, 2.12 GPa and 3.54 GPa. Post processing this data using the IPL-Weibull model yields the plot given in Fig. 4.8. The Weibull shape factor fit to this data is shown in the upper right corner as $\beta = 2.54$. Historically, a shape factor in the range $1 < \beta < 4$ suggests normal bearing and gear-contact-type wear. For comparison, a shape factor less than 1 indicates an infantile failure and suggests that something may be wrong with either the test rig, the hardware, or the testing process. Confidence bands for stress-life curves for 90 % confidence level are plotted in Fig. 4.8 as well.

Table 4.4 Stress cycle data for 33 RCF tests, including suspensions and failures for configuration 1 of Table 4.1

Fail or suspend	Stress cycles	Stress (GPa)	Fail or suspend	Stress cycles	Stress (GPa)
S	3.91E+06	2.12	F	3.03E+07	2.12
S	5.59E+06	2.12	F	3.26E+07	2.12
F	1.02E+07	2.12	F	3.33E+07	2.12
F	1.14E+07	2.12	F	3.35E+07	2.12
S	1.29E+07	2.12	S	3.55E+07	2.12
S	1.49E+07	2.12	F	3.73E+07	2.12
F	1.56E+07	2.12	F	3.94E+07	2.12
S	1.60E+07	2.12	F	4.11E+07	2.12
S	1.68E+07	2.12	F	4.25E+07	2.12
F	1.92E+07	2.12	F	5.22E+07	2.12
F	2.24E+07	2.12	F	7.08E+07	2.12
F	2.24E+07	2.12	F	5.03E+06	3.54
F	2.24E+07	2.12	F	6.34E+06	3.54
F	2.28E+07	2.12	F	7.08E+06	3.54
F	2.28E+07	2.12	F	9.13E+06	3.54
F	2.85E+07	2.12	F	1.02E+07	3.54
F	2.98E+07	2.12			

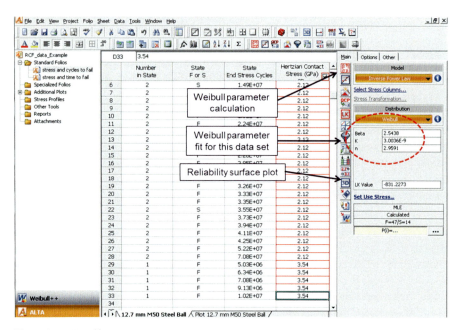

Fig. 4.6 Weibull[++] data screen, with options and choices of model and distribution type

4.3 Rolling Contact Fatigue Vacuum Test

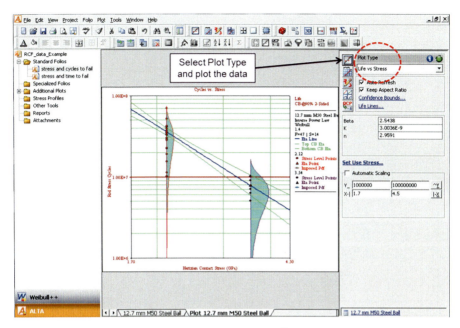

Fig. 4.7 Cycles versus stress plot generated using Weibull++ and the data in Table 4.4

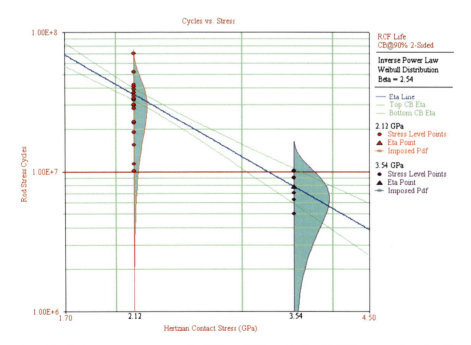

Fig. 4.8 Cycles versus contact stress for 12.7 mm M50 steel balls with M50 races against a Rex20 rod. Test rotation of 130 Hz in high vacuum with approximately 200 nm of silver on the balls (Reproduced with permission from Wear, Volumes 274–275, 27 January 2012, Pages 368–376, Elsevier)

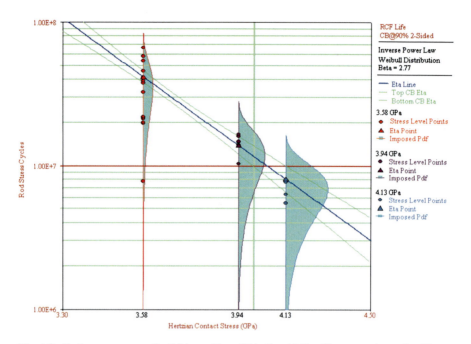

Fig. 4.9 Cycles versus stress for 7.94 mm Koyo T5 balls with Rex20 races against a Rex20 rod. Test rotation of 130 Hz in high vacuum with approximately 150 nm of silver on the balls (Reproduced with permission from Wear, Volumes 274–275, 27 January 2012, Pages 368–376, Elsevier)

It is a good practice to periodically fit the RCF data to the Weibull distribution and compare shape factors over time. As the data is collected, that is, as we continue to accumulate completed tests, it is good to combine old and new data and recalculate the shape factor. Trending the shape factor over time will help identify process or material aberration. A β-shift in the results suggests that the test or the hardware has changed. One may expect a shift from one material lot to the next, but unexpected shifts may reveal lurking variables not accounted for in the statistical analysis.

4.3.5 RCF Test Results

Cycles to failure versus contact stress and reliability for all test configurations in Table 4.1 are presented in this section. Seventy-nine RCF tests were carried out for two ball sizes and two rod materials. Figures 4.8, 4.9, and 4.10 contain cycle versus contact stress for three test configurations presented in Table 4.1. The Weibull shape factor for all test configurations is about 2.6 and is presented in the upper right of each figure. An eta line has been added to each figure for extrapolation for testing at stress levels between the data presented in the figures.

4.3 Rolling Contact Fatigue Vacuum Test

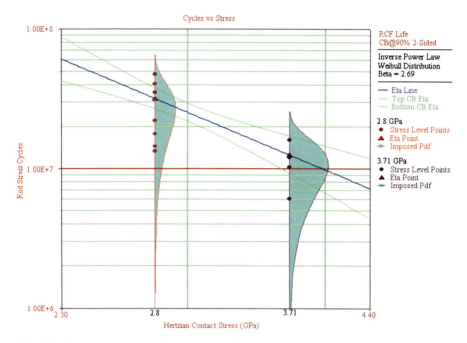

Fig. 4.10 Cycles versus stress for 12.7 mm M50 steel balls with M50 races against a Si_3N_4 rod. Test rotation of 130 Hz in high vacuum with approximately 200 nm of silver on the balls (Reproduced with permission from Wear, Volumes 274–275, 27 January 2012, Pages 368–376, Elsevier)

The shaded area in each figure represents the fitted probability density function based on the RCF data at each stress level. The results in Figs. 4.8, 4.9, and 4.10 strongly suggest rolling contact fatigue failure since the contact stresses were 1/3 less than the calculated tensile yield strength of each component as calculated from the hardness measurements presented in Table 4.2. The RCF life data was used to calculate reliability versus cycles and contact stress, as shown in Figs. 4.8, 4.9, 4.11, and 4.12. A comparison of reliability surfaces reveals that configuration 2 of Table 4.1 allows testing at higher contact stresses than using configuration 1. Reliability surfaces are helpful for planning test length and load for a given hardware configuration. For example, if testing at 2.5 GPa for 10 million cycles is needed, then configuration 2 should be used. If testing at less than 1.0 GPa for one million cycles, then either configuration 1 or 2 may be used, but configuration 1 is less expensive to operate.

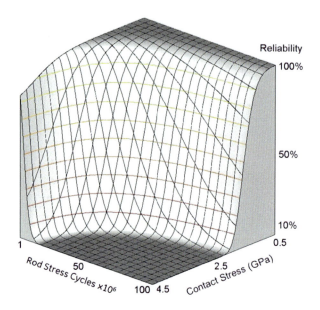

Fig. 4.11 Reliability surface for RCF data in Table 4.4, configuration 1 of Table 4.1, 12.7 mm M50 steel balls with M50 races against a Rex20 rod. Test rotation of 130 Hz in high vacuum with approximately 200 nm of silver on the balls

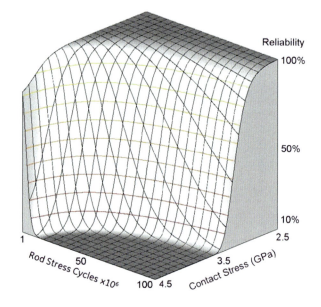

Fig. 4.12 Reliability surface testing with configuration 2 of Table 4.1, 7.94 mm balls with Rex20 races against a Rex20 rod. Test rotation of 130 Hz in high vacuum with approximately 150 nm of silver on the balls

4.3.6 Post-Test Autopsy of Contacting Elements

Post-test autopsy of the balls and rod for each test configuration suggest that lubrication depletion and coating spall hastened the onset of subsurface fatigue failure. Figures 4.13, 4.14, 4.15, 4.16, 4.17, and 4.18 present post-test photos of the

4.3 Rolling Contact Fatigue Vacuum Test

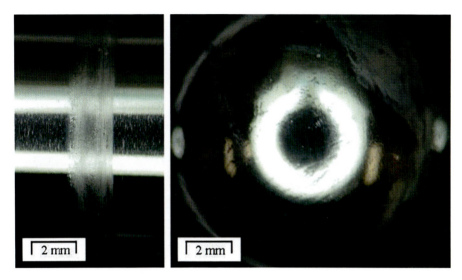

Fig. 4.13 Silver coating spall failure after 5.5 h of testing in configuration 1 of Table 4.1

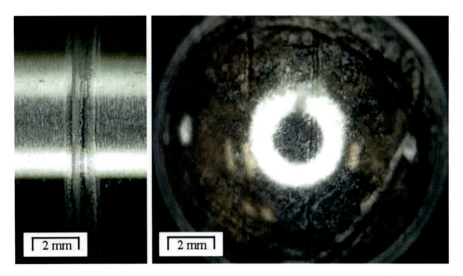

Fig. 4.14 Silver depletion failure after 21.4 h of testing in configuration 1 of Table 4.1

ball and rod for two types of failures, early spall and silver depletion. The test of Fig. 4.13 corresponds to failure due to early spall of the silver coating after 5.5 h. The wear track on the rod in Fig. 4.13 is about twice as wide as the wear track on the rod of Fig. 4.14, which failed due to silver depletion after 21.4 h. Once a coating spall occurred on the surface of at least one of the balls, the surface was damaged due to plastic yielding and the resulting contact area with the rod increased rapidly.

74   4 Rolling Contact Fatigue in High Vacuum

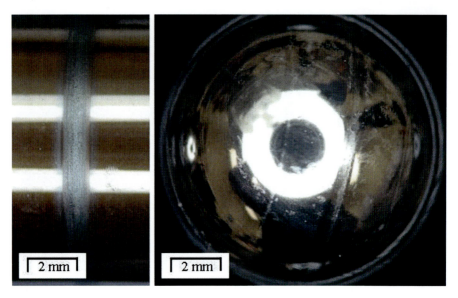

Fig. 4.15 Silver depletion failure after 25.5 h of testing in configuration 3 of Table 4.1

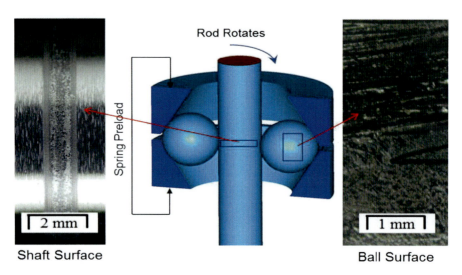

Fig. 4.16 Autopsy data for RCF elements in configuration 1 of Table 4.1 operated at 3.61 GPa loading. Suspension after 1.7 h of rotation at 130 Hz, accounting for approximately 7.8×10^5 rod stress cycles

If vibration in excess of 0.35 g is exceeded for 1 min, the test is stopped and is classified as failed. With the rod rotating at 130 Hz and a deceleration time of about 3 s, approximately 390 rod rotations will be added to the failed surfaces before the rod stops rotating after the stopping criteria have been exceeded.

4.3 Rolling Contact Fatigue Vacuum Test

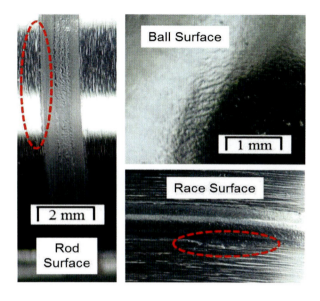

Fig. 4.17 Autopsy data for RCF elements in configuration 1 of Table 4.1 operated at 3.61 GPa loading. Silver depletion failure after 20.6 h of rotation at 130 Hz, accounting for approximately 9.6×10^6 rod stress cycles

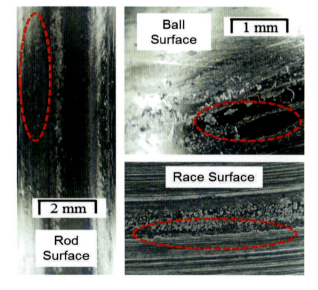

Fig. 4.18 Autopsy data for RCF elements in configuration 1 of Table 4.1 operated at 4.0 GPa loading. Coating spall failure after 9.1 h of rotation at 130 Hz, accounting for approximately 4.2×10^6 rod stress cycles

In comparison, the tests of Figs. 4.14 and 4.15 both failed due to silver depletion. These tests ran out of silver lubrication, and the resulting vibration from non-lubricated contact exceeded the stopping threshold, and the test was stopped before total failure. The wear tracks on the rods in Figs. 4.14 and 4.15 are narrower than in Fig. 4.13. Post-test examination indicates that the silver is pushed out of the wear track on both the rod and race as the test proceeds. The significance of comparing these failures relates to the friction due to incipient wear associated with each type of failure. Spall failure resulted in higher friction, leading ultimately to surface yielding. Silver depletion resulted in increased friction and vibration as well, but without surface yielding before the test was stopped. If allowed to continue without lubrication, the increased friction would accelerate the onset of subsurface spall of the ball and rod.

Figures 4.16, 4.17, and 4.18 reveal interaction between the balls, the rod, and races, at different stages during testing. Post-test autopsy of the rotating parts is extremely valuable and may be used to guide future testing. Mechanisms of wear and material transfer may be investigated for the purpose of validating a model or analysis technique to the test data. For example, the test in Fig. 4.16 was autopsied after just 1.7 h, and results confirm that the elements were in the full-film lubrication regime. There is sufficient silver film present such that asperity-to-asperity contact is less significant. The results in Fig. 4.17 show evidence of mixed-boundary conditions on the contact surfaces. Some of the silver film has been pushed out of the wear track, rendering it useless to the rotating parts. In fact, the film in Fig. 4.17 is very near total depletion, indicating that if allowed to continue the contact would begin entry in the boundary layer regime and rapid failure. Figure 4.18 illustrates a comparison of contacting elements after significant spall failure has occurred. The test failed after 9.1 h, or about 4.2 million stress cycles with 4.0 GPa contact stress. It is unclear which failed first, the silver film or surface material of the ball. However, since the same silver film was used at 4.13 GPa contact stress in configuration 2, and as shown in Fig. 4.9, it is likely that the spall initiated at the ball-coating interface and that the silver film failed for adhesion due to ball surface yielding.

4.3.7 Post-Test Elemental Content of Film

Post-test chemical analysis of the film on the balls may be used to approximate the amount of silver film remaining at the time the test was suspended or failed. Table 4.5 and Fig. 4.19 present scanning electron microscopy (SEM) results of the ball surface for three conditions: before application of silver, after silver depletion failure, and after early failure caused by silver-coating spall. The surface composition of the balls was derived from the energy dispersive spectroscopy analysis attached to the SEM instrument.

Observation of the post-test results in Table 4.5, data from the 25.5 h test in column 3 show no silver present, based on energy dispersion using the SEM instrument. After 25.5 h of rotation or approximately 12.2 million cycles, all of the silver from the ball had transferred to the rod and race in a similar mechanism to the third-body concept used by Higgs and Wornyoh (2008). In comparison, the results of column 4 of Table 4.5 show that silver remains on the ball surface after

4.3 Rolling Contact Fatigue Vacuum Test

Table 4.5 Element composition of one non-coated ball and two coated balls after testing Composition derived from energy dispersive spectroscopy of SEM

	Baseline M50 bearing steel	SEM 12.7 mm M50 ball after 25.5 h, silver depletion	SEM 12.7 mm M50 ball after 7.2 h, early spall failure
Carbon	0.85		
Chromium	4.10	1.48	2.17
Iron	89.30	98.08	80.04
Manganese	0.30		
Molybdenum	4.25		
Silicon	0.20	0.45	0.66
Vanadium	1.00		
Silver	0.00		17.12

Reproduced with permission from Wear, Volumes 274–275, 27 January 2012, Pages 368–376, Elsevier

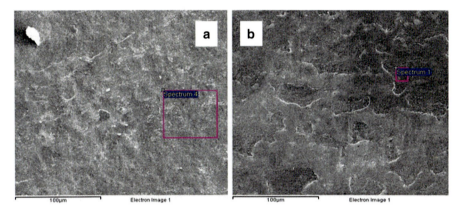

Fig. 4.19 SEM surface of M50 balls from two tests: (**a**) 25.5 h to failure for silver depletion corresponding to column 3 of Table 4.5, (**b**) 7.2 h to early failure for spall corresponding to column 4 of Table 4.5. The "islands" of silver are shown in the darker areas of figure (**b**)

7.2 h and corresponds to an early-life spall failure of the silver coating. Figure 4.19 presents SEM images of the two ball surfaces: (a) In the left panel of Fig. 4.19, silver is depleted from the 25.5 h test, while (b) in the right panel of Fig. 4.19, the silver is present after 7.2 h corresponding to the early spall failure condition. The light and dark color regions highlighted in the right panel of Fig. 4.19 show that silver remains after 7.2 h of testing. The surface texture between the two balls is further indication that the surface in the left panel has little to no silver present. The silver present on the ball surface in the right panel of Fig. 4.19 may be represented as the source input for a third-body transfer mechanism that we will explore in the next section.

4.4 Friction and Wear Calculations

Test life comparisons using two modeling approaches are the focus of this section. The first model uses a conservation of mass approach and is based on the work of Higgs and Wornyoh (2008). A third-body mass transfer concept is applied to account for the transport of the film from the ball surface to the rod and cup contact surfaces as seen experimentally in Figs. 4.16 and 4.17. The second modeling approach is similar to the Lundberg–Palmgren model in that the RCF data from the test configurations of Table 4.1 are used to fit a load-capacity parameter, C, to L_{10} data similar to that found in Chap. 8 of Bhushan (1999).

4.4.1 Third-Body Transfer Mechanism

The model developed in this section follows the work of Higgs and Wornyoh (2008) and has been extended to the ball–rod–cup contact of the RCF platform. The third-body transfer scheme is based on the assumption that the solid lubricant will transfer or smear onto any surface that it contacts. This smearing or transfer mechanism has been observed for testing in configuration 4 of Table 4.1 using a bronze carrier cage. Within 20 min of testing, the balls were bronze in color, and all of the silver initially present on the balls had become mixed with the bronze material. For the study described in this section, no carrier cage is used, and the coating on the surface of the ball represents the lubrication supply, approximately 200 nm of silver coating. The silver coating on the ball is analogous to the pellet supply in Higgs and Wornyoh (2008). The third-body storage volumes are located in the valleys of the asperity within the wear track of the cups and the rod. The control volume fraction coverage model (CVFC) has been presented and explained in Higgs and Wornyoh (2008); however, some parts of that formulation will be presented here for clarity. The assumptions of the CVFC model for solid lubrication transfer to the third-body are as follows: (i) the ball–rod and ball–cup contact surfaces are flat within their contact areas, (ii) incipient sliding occurs between surfaces due to elastic deformation, and (iii) the fractional response and friction of the interfaces is primarily a function of the amount of silver present in the third-body volumes of the cup and rod and on the surface of the ball.

The conservation of mass formulation for the transfer of silver to the wear tracks of the race and rod is given as

$$\begin{pmatrix} \text{Third Body} \\ \text{Storage Rate} \end{pmatrix} = \begin{pmatrix} \text{Third Body} \\ \text{Input Rate} \end{pmatrix} - \begin{pmatrix} \text{Third Body} \\ \text{Output Rate} \end{pmatrix}. \quad (4.9)$$

The output rate in Eq. 4.9 is driven by the load between the ball–rod and ball–cup that forces some of the solid silver out of the wear track. Examination of the wear tracks on the cups and on the rod in Figs. 4.16, 4.17, and 4.18 reveals that silver is

4.4 Friction and Wear Calculations

pushed outside of the CVFC volume over time and hence removed from the third-body storage volume. The input rate to the third-body storage volumes on the rod and race contact areas is similar to that presented in Higgs and Wornyoh (2008) in that silver is pushed into the valleys between surface asperities within the contact area of the rod and cup. Incipient sliding between the ball–rod and ball–cup is assumed throughout the formulation.

The fractional coverage $X(t)$ of the third bodies on both the rod and the cup, or race wear tracks, is normalized to the roughness of the race and rod surfaces, approximately 0.25 µm, and represents the maximum asperity height, h_{max}. The asperity depth is about the same as the initial silver coating thickness on the balls as well, approximately 0.21 µm. Following the form of Higgs and Wornyoh (2008), the fractional coverage variable is defined as

$$X = \frac{h}{h_{max}}, \tag{4.10}$$

where h is the local height of the silver film in the third-body volumes. Archard's volume wear rate law is used to account for surface wear interactions and is defined as

$$\frac{dV}{dt} = KF_N U, \tag{4.11}$$

where V, K, F_N, and U are the volume, wear coefficient, normal force, and sliding velocity, respectively. The wear coefficient K is the probability that a surface is being worn due to sliding contact, and for this section, incipient sliding has been assumed. Combining Eqs. 4.9, 4.10, and 4.11 gives the following differential equation for $X(t)$ as

$$Ah_{max}\frac{dX}{dt} = (K_{bc}F_c U_c + K_{br}F_r U_r)(1 - X) - (K_{bEc}F_c U_c + K_{bEr}F_r U_r)X, \tag{4.12}$$

where the first term on the right-hand side accounts for third-body input and the second term for third-body removal. Equation 4.12 is the wear volume rate and the constant A is the area of contact. The solution of Eq. 4.12 is given as

$$X(t) = \frac{K_{bc}F_c U_c + K_{br}F_r U_r}{K_{bc}F_c U_c + K_{br}F_r U_r + K_{bEc}F_c U_c + K_{bEr}F_r U_r}\left(1 - \exp\left(-\frac{t}{\tau}\right)\right) \tag{4.13}$$

The constants K_{bc} and K_{br} are the wear coefficients for silver between the ball–race and the ball–rod, respectively, and influence how the third body is filled with silver from incipient sliding during the test. The constants K_{bEc} and K_{bEr} are the wear coefficients for the silver that is pushed out of the wear track between the ball–race and ball–rod. The wear coefficients K_{bEc} and K_{bEr} influence how much silver is removed from the third body due to ball contact with the edges of the wear track.

Table 4.6 Wear coefficients used in Eqs. 4.13 and 4.14

K (m^2/N)	K$_{bc}$	K$_{br}$	K$_{bEc}$	K$_{bEr}$
Test configuration 3: Si$_3$N$_4$ Rod, M50 ball and race	1.0E-15	2.0E-16	1.0E-15	5.0E-17
Test configuration 1: Rex20 Rod, M50 ball and race	1.0E-15	2.0E-15	1.0E-15	2.0E-17

Reproduced with permission from Wear, Volumes 274–275, 27 January 2012, Pages 368–376, Elsevier

Table 4.7 Normal force and third-body contact area calculations

Hertz contact stress (GPa)		F$_r$ (N)	F$_c$ (N)	Third-body surface area (m^2)
Test configuration 3: Si$_3$N$_4$ Rod, M50 ball and race	3.7	264.1	145.2	7.12E-05
	2.8	101.8	56.1	5.37E-05
Test configuration 1: Rex20 Rod, M50 ball and race	3.5	237.6	130.7	6.10E-05
	2.2	67.9	37.3	3.88E-05

Reproduced with permission from Wear, Volumes 274–275, 27 January 2012, Pages 368–376, Elsevier

The edges of the wear track on the race are illustrated in Fig. 4.17 and are considered silver lubricant removal mechanisms during rolling contact. The time constant τ in Eq. 4.13 is defined as

$$\tau = \frac{Ah_{max}}{K_{bc}F_cU_c + K_{br}F_rU_r + K_{bEc}F_cU_c + K_{bEr}F_rU_r} \quad (4.14)$$

and defines the time to steady-state third-body film thickness. It was found that τ also correlates with the run-in time of the RCF test configurations in Table 4.1. The condition $X(t) > 0$ signifies that silver is being transferred from the ball surface to the third-body volumes on the race and rod. When all of the silver has been transferred from the ball, the condition $X(t) = 1$ exists and the third-body input rate goes to zero as defined in Eq. 4.12. As the third-body volume becomes depleted, that is, as silver is pushed out of the wear track as defined in the second term on the right-hand side of Eq. 4.12, the test result of Table 4.5 column 3 begins to occur and highlighted in Fig. 4.17 as well.

As soon as the input to the third-body volume ceases, the third-body volume coverage $X(t)$ diminishes resulting in asperity-to-asperity contact, similar to the boundary layer contact described in Sect. 4.1, such that friction and vibration increase and the stopping vibration threshold for the RCF test is exceeded. Table 4.6 presents the wear coefficients used in Eqs. 4.13 and 4.14. These values of wear coefficients are within the range and order of magnitude of those tested between bearing steels like Rex20 and silver in high vacuum as seen in NASA/TM (1999) and those tested between Si$_3$N$_4$ and silver in Holmberg and Matthews (2009). Table 4.7 contains normal load and contact area calculations from the RCF test

4.4 Friction and Wear Calculations

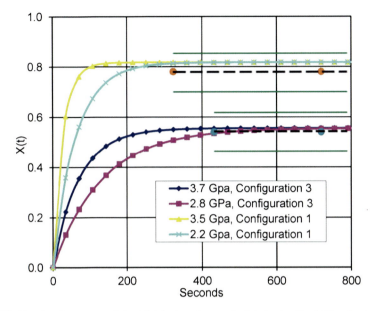

Fig. 4.20 Fractional coverage of the third-body volume calculated using Eq. 4.13 and values from Tables 4.2, 4.6, and 4.7 for two test configurations from Table 4.1

setup shown in Fig. 4.5. Equation 4.13 is plotted in Fig. 4.20 using the material properties, wear coefficients, and loads presented Tables 4.2, 4.6, and 4.7.

Post-test thickness measurements of the silver remaining on the balls suggests a third-body coverage steady-state value, X_{SS}, between 0.46 and 0.57 when testing in configuration 3 and X_{SS} between 0.72 and 0.88 when testing in configuration 1. These values represent the measured steady-state fractional coverage of the third-body without spall failure and before coating depletion. X_{SS} was computed similar to Eq. 4.10 in that the measured silver thickness on the ball was normalized using asperity height, h_{max}. The measured steady-state coverage X_{SS} and the calculated coverage $X(t)$ are plotted in Fig. 4.20. There is agreement between measured and predicted third-body fractional coverage using the wear coefficient values presented in Table 4.6.

Observation of Fig. 4.20 reveals the trending of $X(t)$ to a steady-state value for each of the configurations 1 and 3. The steady-state value of $X(t)$ is dependent on the material type and loading conditions related to the RCF test setup. There is equal force applied to the ball, race, and rod surfaces, and this condition will drive $X(t)$ to a constant value. For comparison, Higgs and Wornyoh (2008) showed trending to steady-state coverage $X(t)$ was dependent on the ratio of the source pellet to the slider loads.

The run-in time for each test configuration 1 and 3 is comparable with the transient portion of the curves in Fig. 4.20. During the first 10–20 min of the RCF test, the vibration peaks to about 0.3 g, representing the run-in portion of the test.

Table 4.8 Steady-state wear factor of the ball, calculated using data from non-spall RCF tests

Table 1 test configuration	Configuration 1, 12.7 mm M50 ball, 9.53 mm Rex20 rod, M50 races		Configuration 2, 7.94 mm T5 ball, 9.53 mm Rex20 rod, Rex20 races		Configuration 3, 12.7 mm M50 ball, 9.53 mm Si$_3$N$_4$ rod, M50 races	
Contact stress (GPa)	3.54	2.16	4.14	3.58	3.71	2.81
Wear factor (cm^{3}cm^{-1}kg^{-1})	3.38E-10	3.28E-10	7.07E-11	3.24E-11	3.13E-10	2.31E-10

Reproduced with permission from Wear, Volumes 274–275, 27 January 2012, Pages 368–376, Elsevier

The increase of $X(t)$ in Fig. 4.20 correlates with the run-in time observed during testing and suggests that the volumes between asperities on the rod and races fill-up within the first 10 min of the test rotating at 130 Hz.

The steady-state wear factor for the depletion of silver from the ball may be calculated using Archard's wear equation integrated over time as

$$V_{\text{ball}} = \int_0^{t_f} K_{\text{ball}} F_{\text{ball}} U_{\text{ball}} (1 - X(t)) dt. \quad (4.15)$$

Integration of Eq. 4.15 with use of Eq. 4.13 followed by rearrangement gives the ball steady-state wear factor as

$$\varphi = \frac{V_{\text{ball}}}{F_{\text{ball}} t_f U_{\text{ball}}} g_r, \quad (4.16)$$

where g_r is the gravitational constant and t_f is the time to failure based on the stopping threshold criteria 0.35 g. Table 4.8 contains evaluation of Eq. 4.16 using data from tests presented in Figs. 4.8, 4.9, and 4.10, excluding early spall failures. Configuration 2 shows the smallest wear factor and had the longest RCF test life. The wear factors of configurations 1 and 3 are about the same, suggesting similar test-time results using either the Rex20 rod or the Si$_3$N$_4$ rod with 12.7 mm balls. Similar to Figs. 4.11 and 4.12, the wear factors calculated in Table 4.8 may be used when one is planning for the next set of RCF tests.

4.4.2 Empirical Comparison: Lundberg–Palmgren Model

A calculation of L_{10} life is presented next and follows Bhushan (1999), Chap. 8. The stress cycles corresponding to 10 % failure may be calculated as

4.4 Friction and Wear Calculations

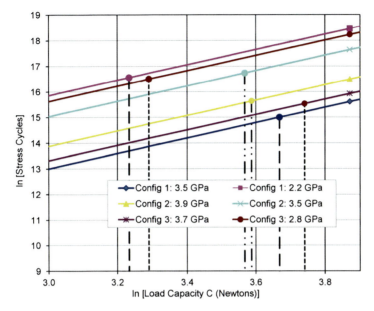

Fig. 4.21 Natural log plot of basic load capacity using Eq. 4.17 and the ball–rod radial loads for configurations 1 through 3 defined in Table 4.1. L_{10} life experimental results also plotted as vertical lines corresponding to RCF stress cycle life and calculated load factor

$$L_{10} = \left(\frac{C}{W}\right)^3, \tag{4.17}$$

where the variable W corresponds to the radial load applied to the ball and C is the basic load capacity of the test configuration with respect to a ball bearing-type system. The basic load capacity C may be calculated using the RCF test results presented in Figs. 4.8, 4.9, and 4.10. The values L_{10} and W are calculated and defined for each test configuration from Fig. 4.1. Figure 4.21 presents basic load capacity calculations for the test configurations presented in Table 4.1. Also plotted in Fig. 4.21 are the stress cycle values based on the L_{10} life from the RCF experiments of Figs. 4.8, 4.9, and 4.10, represented as vertical lines of a specific type (dashed, hashed, dotted).

It is interesting that configurations 1 and 3 give different load capacities for the same configurations, suggesting that life-adjustment factors are needed to accurately calculate L_{10} life for these test element combinations. It is to be expected that the load capacity will be constant for each test configuration independent of bearing load. The load capacity related to configuration 2 is similar for both load magnitudes, suggesting that Eq. 4.17 alone may be used to predict test life based on load, W, and load capacity, C, for this configuration.

Conclusions and Observations

The ball–rod–cup RCF test platform has been validated for testing thin solid film lubricants in high vacuum and at high rotational speeds. Post-test results indicate that film thickness initially present on the balls has the most influence over test life based on 118 tests and a thickness range of 180–200 nm. The Weibull shape factor for all test data was in the range of 2.5–2.8, indicating typical bearing-type failure modes and test times. A shape factor in this range suggests predictability of the test method, which enables repeatable conclusions concerning tribology testing of rolling bearing elements.

For the two ball sizes investigated in this chapter, the 7.94 mm ANSI T5 gave longer RCF test life than did the 12.7 mm M50 steel balls even for higher Hertz contact stress loading. Surface finish of the balls and races also influenced test longevity, with longer life and higher load capability for the 7.94 mm T5 balls with a surface Ra of 0.04 compared to the 12.7 mm M50 balls with a surface Ra of 0.32. The surface roughness of the races directly affects the volume of silver that may be stored within the wear track over the course of the test. Rod material and surface finish did not influence test time for rod materials Rex 20 steel and Si_3N_4.

Two types of failure modes were observed for all tests: (i) early spall failure and (ii) silver film depletion. In the former, a film spall on at least one ball resulted in increased friction and plastic yielding within the contact area, leading to the onset of spall at the ball film interface. In the latter failure mode, as the lubricant film became depleted, the friction and vibration increased and the test was stopped. Surface finish of the rod and race elements influences the rate of solid film transfer from the ball surface to the rod and races. Inspection of the wear surfaces of suspended test components suggested a third-body transfer mechanism of the lubricating film could be used to predict test longevity. A third-body self-replenishing model approach was applied, and there is good agreement with measured steady-state film thickness and predicted film height.

A Lundberg–Palmgren empirical model fit of the RCF test data for L_{10} life was used to calculate the load-capacity parameter for three test configurations. Stable prediction of a load capacity parameter will enable better planning for expanded used of the high-vacuum RCF test platform. The load capacity parameters for configurations involving 12.7 mm balls with M50 races varied with applied load, suggesting that further work is needed to establish load-reduction correction factors for these configurations. Load-reduction factors for configurations involving 7.94 mm Koyo T5 balls with silver film are not required. Now that this test platform has been validated, testing for influence of process parameters on RCF test life is investigated in Chap. 6.

References

Bhushan B. Principles and applications of tribology. New York: Wiley; 1999.

Danyluk M, Dhingra A. Rolling contact fatigue using solid thin film lubrication. Wear. 2012a;274–275:368–76.

Higgs CF, Wornyoh EYA. An in situ mechanism for self-replenishing powder transfer films: experiments and modeling. Wear. 2008;264:131–8.

Holmberg K, Mathews A. Coatings tribology: properties, techniques and applications in surface engineering. Danvers: Elsevier; 2009.

Hoo J. A ball-rod rolling contact fatigue tester. In: ASTM, editor. ASTM STP 771. Baltimore: ASTM; 1982. p. 107–24.

Hoo JC. STP771: rolling contact fatigue testing of bearing steels. Phoenix: ASTM; 1981.

Hoo JC. STP987: effect of steel manufacturing processes on the quality of bearing steels. Philadelphia: ASTM; 1988.

Mattox D. Handbook of physical vapor deposition processing. Westwood: Noyes; 1998.

NASA1999-209088. Friction and wear properties of selected solid lubricating films. Part 1. Bonded and magnetron-sputtered molybdenum disulfide and ion-plated silver films. Technical report, NASA; 1999.

Sadeghi F, Jalalahmadi B, Slack TS, Raje N, Arakere NK. A review of rolling contact fatigue. J Tribol. 2009;131:1–15.

Totten G, Liang H. Surface modification and mechanisms: friction, stress, and reaction engineering. New York: Marcel Dekker; 2004.

Chapter 5
Coating Thickness Calculation and Adhesion

Abbreviations

t_h	Calculated film coating thickness
ρ_{Ag}	Density of high-purity silver
AES	Auger electron spectroscopy
RCF	Rolling contact fatigue
Si_3N_4	Silicon nitride material
TiN	Titanium nitride
XRF	X-ray fluorescence spectroscopy

Coating thickness measurement over curved surfaces such as ball bearings and circular cross section rods is difficult. One could attempt to measure thickness directly by probing the coating material with a beam of electrons, starting from the coating surface and penetrating by a sputter method through to the surface of the ball. Probe-based methods rely on spectroscopy tools and may require some surface preparation before measurement. For example, Auger electron spectroscopy (AES) requires high-vacuum conditions to enable an electron beam to knock-off film atom electrons as the beam penetrates the film surface. The energy emission resulting from the shift in electron shells is characteristic to the film material and may be compared with base standards to confirm the film composition. Once the probe is complete, a sputter technique is used to remove a layer of atoms and the process is repeated through the film. The AES method, as well as others that use spectrometry, rely on sputtering-rate curves of a similar material of known thickness and sputtering time. Typically a surface chemistry-sensitive method such as X-ray photoelectron spectroscopy (XPS) is needed to help interpret the AES results.

Some spectroscopy methods require less sample preparation compared to AES and XPS. Glow discharge optical emission spectroscopy (GDOES) is an optical emission method that may be used to quantitatively detect element type in thin films. GDOES requires less sample preparation than AES and does not need ultrahigh-vacuum conditions. A glow discharge is used to bore through the coating as the spectrometer tracks atom removal and a sputter rate is assumed for the film. Thickness is calculated as the product of the sputter rate and total time to reach the base material, in our case the surface of the ball or rod. The film thickness measurement is completed when the spectrometer detects atoms of the base material.

Film adhesion may be measured using scratch testing, as well as by visual inspection based on a sampling scheme for a given quantity of coated balls. For example, a stylus or probe tip applies a force to the film surface while it is moving, and the system tracks the magnitude of resistance to motion. Also, an indentation test may be used to quantify specific characteristics of the film such as hardness and response to deformation. In this chapter we discuss two nondestructive film thickness measurements used to quantitatively calculate film thickness on silver-coated ball bearings. We also discuss low-cost and low-overhead methods to assess film adhesion. The intent of this chapter is to give the reader an introductory overview of thickness and adhesion measurement tools. A more detailed discussion may be found in Totten and Liang (2004) and Ohring (2001).

5.1 Thickness Measurement Techniques

Coating thickness measurement by weight is an inexpensive and nondestructive method to determine film thickness on the surface of ball bearings. In this method, the balls are weighed before and after coating and a uniform thickness over all ball surfaces is assumed. This method works well for simple shapes such as ball bearings and rods, but it would not work well for complex shapes or over surfaces of known thickness variation. For example, consider application of a coating to the major and minor diameters of an M8 bolt fastener threads. The coating thickness will vary between the major and minor diameters of the threads and therefore the weight measurement method will not accurate for this application. Calculating coating thickness based on weight is nondestructive and allows one to test and then use the coated balls after measurement. In contrast, scratch testing methods and the AES method permanently damage the coating during measurement and are classified as destructive methods.

Accuracy of the weight measurement method increases with the number of balls included in the measurement. This method works well for large numbers of balls coated in a single process. For the balls coated in Chaps. 6 and 7, the coating lots ranged from 60 to 320 balls in a single coating process. Further, for soft lubricating films such as silver, small variations in thickness on the ball surface will be mitigated by rolling contact during run-in. In comparison, thickness variation in hard films such as TiN applied to ball bearings would result in nonuniform stresses

5.1.1 Calculate Thickness by Weight

Calculating coating thickness by weight is based on the assumption of a uniform film thickness over all ball surfaces. With this assumption, one may compare pre- and post-coated ball weight and then calculate an average film thickness based on the weight of the film. The density of the film material, silver, for example, ρ_{Ag}, and ball diameter are all that is needed for the thickness calculation. Accuracy improves when the coating weight is averaged over a large number balls. For example, 60 balls will give more accurate average-thickness calculation than would 2 or 3 balls average thickness.

Example 5.1: Ball Coating Thickness Calculation by Weight Method Consider a thickness calculation based on ball and coating weight. The balls are ANSI T5 material and 7.94 mm in diameter. Sixty uncoated balls were weighed to within four significant decimals, as 135.8299 g. The post-coating weight of these 60 balls is 135.8411 g.

Solution: The surface area of one ball is calculated as

$$A_s = 4\pi R^2 = 197.9324 \text{ mm}^2. \tag{5.1}$$

The coating density is given as

$$\rho_{Ag} = 0.01049 \text{ }^g/_{mm^3} \tag{5.2}$$

The weight of the coating for all 60 balls is (135.8411–135.8299) g or 0.0112 g. The coating weight per ball is 0.0112 g divided by 60 or 1.87×10^{-4} g. The coating weight per ball is related to coating density and coating volume, as

$$\rho_{Ag} 4\pi R^2 t_h = 1.87 \times 10^{-4}. \tag{5.3}$$

Solving Eq. 5.3 for thickness t_h and using Eqs. 5.1 and 5.2, the coating thickness is 9.0×10^{-5} mm, or 90.0 nm. This calculation is presented in Table 5.1 using an Excel Spreadsheet. When using a spreadsheet calculation or other numerical code, retain at least six decimal digits in the calculation to improve accuracy.

For the calculation in Example 5.1, the assumption of uniform thickness is sufficient considering the way in which these balls are coated with silver. Figure 5.1 shows coated balls in a carousel after coating. The balls are bathed in silver vapor and argon plasma and are randomly circulated within the pockets of the carousel

Table 5.1 Silver coating thickness calculation by weight

Ball diameter	7.937500	mm
Coating material	Silver	
Coating density	10.49	gram/cc
Ball surface area	197.932609	mm^2
Multiple balls		
Number of balls	60	Each
Total weight before coating	135.829950	gram
Total weight after coating	135.841130	gram
Total coating weight	0.011180	gram
Average single ball coating weight	0.000186	gram
Average single ball coating thickness	89.742407	nm

Fig. 5.1 Silver-coated ball bearings: 7.94 and 3.97 mm diameter balls in carousel pockets

during coating. Figure 5.2 illustrates the relative location of the carousel and other elements inside the chamber and in particular the silver source at the bottom of the chamber. Heat is applied to the silver source and the silver is evaporated from the bottom of the chamber. The silver atoms migrate upward to coat the balls near the top of the chamber. The balls in Example 5.1 were coated using an ion plating method in a large chamber as shown in Fig. 5.2. A more detailed description of the Ion plating process and the chamber in Fig. 5.2 is presented in Chap. 6. The balls in the carousel are permitted to roll around to insure uniform coating thickness over all of the balls. The carousel is mechanically agitated to promote ball meandering in the pocket during coating.

5.1 Thickness Measurement Techniques

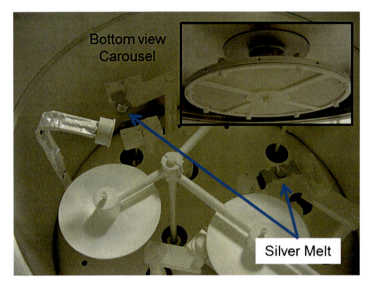

Fig. 5.2 Inside the ion plater chamber with ball carousel in inset

5.1.2 Thickness Measurement Using XRF Spectrometry

The thickness calculation in Sect. 5.1.1 required pre- and post-coating ball weight. A less labor intensive method using X-ray fluorescence (XRF) spectroscopy does not require multiple steps to calculate average thickness. The XRF method is simpler than the method in the previous section since there are no measurements or calculations required by the operator. The XRF method is used extensively in nondestructive chemical analyses of solid substances in bulk form. There is little sample preparation needed for XRF method, and the measurements are taken under vacuum conditions so there is less risk of contamination during handling. In the XRF method, the coated balls are bombarded with intense X-rays while a detector measures the energy levels that are reflected off the coating. The amount and type of X-ray radiation that is reflected off the coated balls is dependent on coating chemistry, composition, and thickness.

The XRF method is used in a comparative manner. The measured X-ray energy scattered off the coating is compared with the energy scattered from a sample with known composition, chemistry, and thickness. One drawback of XRF spectroscopy is that it requires thickness-calibration standards for X-ray energy reference to compare with during coating measurement. One significant advantage of XRF compared to the weight method is that it can detect the presence of multiple elements within the coating. The thickness-by-weight method assumes a uniform composition, all silver in Example 5.1. The XRF method may be used to detect quantities of multiple elements, nickel-copper-silver, for example, while measuring thickness. It is this aspect of XRF that we use in Chap. 6 to confirm coating thickness. A note of caution though, the XRF method cannot detect the thickness of each layer; only the total amount of each element present in the coating is detected. Layer thickness and composition of each layer within a multilayer coating requires destructive investigation of the coating. Auger electron spectroscopy (AES) is well suited for layer-by-layer analysis and is used in Chap. 7.

5.2 Pretest Adhesion Check

A qualitative adhesion test is useful for screening coated balls before installing into the test rig. Examination under microscope of a fixed sample size of balls can give insight into coating performance prior to RCF testing. For example, if there was oxygen present during the coating process, the silver coating will have a different texture compared to a non-contaminated coating when viewed under microscope. If the process voltage was too high, the silver coating may be less malleable and may even show evidence of flaking.

Coating adhesion as well as cohesion may be checked using the ASTM-B905 tape test, ASTM-B905-2010 (2010). Coating adhesion pertains to the interfacial bond strength of the coating to the ball surface. Cohesion relates to bonding strength within the coating layer. ASTM-B905 uses flexible tape with an adhesive of known sticking force to qualitatively try to remove any coating particulate as a result of poor adhesion or cohesion. However, surface preparation of the coated ball is required as outlined in ASTM D3359; Standard Test Methods for Measuring Adhesion by Tape Test. For testing in high vacuum and in rolling contact fatigue, the tape-test methods are considered destructive since the balls may not be used after the ball surface has been contaminated with the carbon-based glue from the tape.

5.2.1 Scratch Test Ball Sample

The malleability of a lubricating film may be checked using a scribe or similar pointed-tip tool while under microscope. Scratching the silver coating and observing its surface response will give insight about coating performance. Figure 5.3 contains photos of a silver-coated, 7.94 mm diameter ball using a camera mounted to medium power microscope. The silver coating is malleable, but there are also globs of silver as a result of the plowing of the probe tip. The inset photo in the lower right of Fig. 5.3 indicates there is good adhesion to the ball surface, but the coating cohesion is poor as suggested by the silver globs. In addition, the coating away from the scratch area is spotty and nonuniform in color. These observations were correlated with oxide contamination in the coating using AES, and in fact this ball was coated in a partial oxygen and argon plasma.

5.2 Pretest Adhesion Check

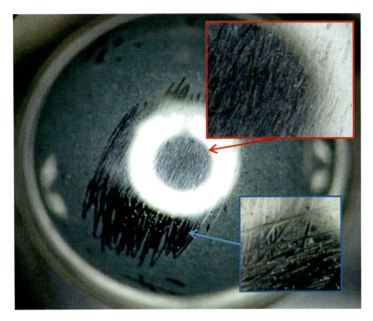

Fig. 5.3 Silver coating on a Si_3N_4 ball of 7.94 mm coated in a 70/30 sccm argon–oxygen plasma at 17 mTorr process pressure in the chamber of Fig. 5.2

The coated surface of Fig. 5.4 has a darker color than in Fig. 5.3, and there are more silver globs on the surface as well. The silver coating is not as malleable either and is thinner than the coating of Fig. 5.3. This film coating was applied in an argon plasma at a higher process voltage compared to Fig. 5.3. For the coating in Fig. 5.4, the problem was not oxygen contamination, rather the silver globs are made up of contaminate metals from the fixtures inside the chamber. The contaminant atoms were sputtered off the stainless steel fixtures as a result of high process voltage during coating. The effects of voltage, pressure, and ball type will be explored in Chaps. 6 and 7. The point of these examples is that one may quantify coating behavior using low-cost and low-overhead tools and observation and then photograph the results for future correlation with RCF life performance.

5.2.2 Particulate Detection Tape Testing

The ASTM 3359 tape test, ASTM-C3359 (1999) is a systematic way to detect and collect coating particulate due to poor film adhesion and cohesion. After the coating debris is collecting on the tape, the particulates may be measured and recorded for future correlation with RCF life test data. For example, Fig. 5.5 contains tape-test results for two titanium nitride (TiN)-coated cylindrical surfaces. For illustration, the TiN coating particulate is easier to see in Fig. 5.5 than for silver particulate. The

Fig. 5.4 Silver coating on Si$_3$N$_4$ ball of 7.94 mm coated in pure argon plasma at high voltage 3.5 kV and at 17 mTorr pressure using the chamber shown in Fig. 5.2

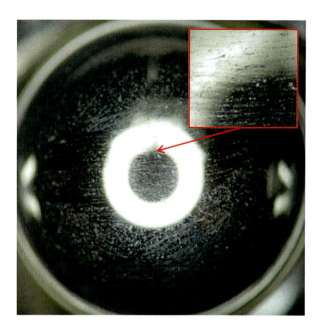

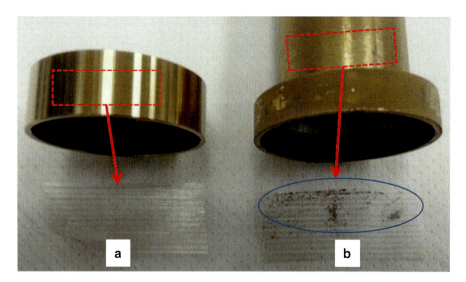

Fig. 5.5 Comparison of tape-test results for good adhesion (**a**) and poor adhesion (**b**) of a TiN coating applied to cylindrical-shaped stainless steel parts

result on the left (a) shows no visible TiN flakes, while the result on the right (b) has large flakes from the coating surface. The TiN coating on the right (b) is heavily oxidized and tape test indicates cohesive failure of the coating. For this example, the TiN flakes are clearly seen without help from a microscope.

Tape test results are sometimes used in conjunction with the XRF spectroscopy presented in Sect. 5.2.1. The failed coating flakes may be too small to detect using a microscope, and XRF is used to quantify the amount of coating that was removed during the test. For example, a tape test applied to ball bearings would require that the entire ball surface make contact with the tape. Simply put, one would roll the ball onto the tape adhesive until all the ball surface has contacted the tape. Then, XRF analysis is used to measure the amount of silver present on the tape. The amount of silver left on the tape is indication of adhesion and cohesion quality.

5.3 Closing Comments

Thickness and adhesion measurements may be accomplished using common laboratory tools and methods. Examination under microscope of a fixed sample size of coated balls might be sufficient as a final check before installing into the RCF test rig. With limited resources, it is better to put time and money into improved deposition tools and chambers rather than defect detection equipment such as AES and XRF. Keep in mind that the methods of this chapter are post-coating investigations and that the damage if detected has already been done. A better solution and a more economical use of resources is to prevent coating damage in the first place. Methods to identify and improve coating quality are the focus of the next four chapters.

References

ASTM-B905-2010. Standard test methods for assessing the adhesion of metallic and inorganic coatings by the mechanized tape test. Technical report, Philadelphia: ASTM; 2010.
ASTM-C3359. Standard test methods for measuring adhesion by tape test. Technical report, New York: ASTM; 1999.
Ohring M. Material science of thin films, deposition and structure. 2nd ed. San Diego: Academic; 2001.
Totten G, Liang H. Surface modification and mechanisms: friction, stress, and reaction engineering. New York: Marcel Dekker; 2004.

Part II
Simulation and Testing of Thin Films in a Vacuum Environment

Chapter 6
Ion-Plating Process Model

Abbreviations

dN/dt	Atomic evaporation rate
V_{sp}	Atomic volume associated with sputter yield
L	Coil inductance
R_l	Coil resistance
q	Current charge
b	Damping coefficient of conductance valve model
m	Element mass
σ_i	Ion collision cross-section coefficient
J	Ion current density
n_i	Ion density
$\bar{u}_i$	Ion group velocity
λ_i	Ion mean-free path
$\Omega_{t,i}$	Kinetic energy of target atoms and incident ions
G	Linear system plant model
J_l	Mass moment of inertia of conductance valve
T_d	PID controller derivative time
K_c	PID controller gain
T_i	PID controller integral time
K_p	Plant gain for thickness monitor plant model
τ	Plant model time constant
n_g	Plasma gas density
s	Plasma sheath thickness
R_{sp}	Sputter removal rate
Y_{sp}	Sputter yield
L_d	System dead time

© Springer International Publishing Switzerland 2015
M. Danyluk, A. Dhingra, *Rolling Contact Fatigue in a Vacuum*,
DOI 10.1007/978-3-319-11930-4_6

p^*	Vapor pressure of liquid silver
φ	View factor for evaporation rate
$K_{mx}\ K_f$	Gas flow inlet system calibration factors
$Q_{in}\ Q_{out}$	Gas flow rate in sccm
BN	Boron nitride
DoE	Design of experiments
LQR	Linear quadratic regulator
PID	Proportional–integral–derivative
RGA	Residual gas analysis
Sccm	Standard cubic centimeters per minute
TiC	Titanium carbide
TiN	Titanium nitride
HV-RCF	High-vacuum rolling contact fatigue
W	Watts unit of power

The goal of this chapter is to model a deposition process and then to compare simulation results with experimental data. With this comparison, a link can be established between model prediction and coating performance. Simulation of vacuum processes and in particular thin-film deposition processes is beneficial for the following reasons: (i) to aid decision making concerning purchase of equipment, (ii) scheduling and allocation of resources for equipment maintenance, and (iii) a quantitative evaluation of product and materials that may have been altered during a process aberration. The first and second of these have been explored extensively over the years. Indeed, production managers and process engineers have developed elaborate tools for cost projection and resources requirements based on customer demand and raw material availability. Many of these tools may be run in Microsoft Excel which further enhances their usefulness, and many are based on a run-to-run comparison philosophy.

Evaluation of the effects of process aberration has received less attention in recent years, partially due to use of lower-cost materials and partially due to improved process control and detection equipment and algorithms that can rapidly correct and adjust the process to maintain the needed level of quality. A lot of work has been done to simulate the micro, nano, and quantum processes associated with thin-film deposition within the last two decades. Numerical simulations have been verified using postdiagnostic tools such as SEM, TEM, and Auger spectroscopy to name a few. But these tools are passive or observational in nature and may not be sufficient to quantify the effects that process aberrations have on the end use of the film. For example, a suspect coating that is applied to ball bearings may pass thickness and adhesion tests, but still fail due to spall once installed in the final application. It would be helpful to measure, test, and quantify the effects of process aberration on coating performance rather than scrap or "reject" a batch of material for process aberration. One could establish a process aberration tolerance and avoid scrapping a potentially good batch of coated balls.

6.1 Plasma and Deposition Processes

In this chapter an ion-plating deposition process is simulated using Simulink along with first-principle deposition and linear control models. The ion-plating process is well suited for deposition of thin films onto large numbers of substrates in a single process for thicknesses in the range of 50–200 nm. The process has been in use since the 1970s and is well understood and was chosen in this chapter to study process aberration effects on film quality. Ion-plating deposition is very dependent on chamber pressure, bias voltage, and bias current within the plasma. The candidate substrates in this chapter are 9.53 and 12.7 mm diameter ball bearings, similar to the ball sizes and materials presented in Chap. 4. The balls are coated with 90–120 nm of silver under aberration conditions. Additionally, a 10 nm layer of nickel (Ni) and copper (Cu) are co-deposited to improve adhesion. The nickel and copper, Ni–Cu, films are sputter-deposited using a combined-material target positioned above the fixture containing the balls. The coated balls are then tested in rolling contact fatigue (RCF) in high vacuum using the test rig of Chap. 4.

6.1 Plasma and Deposition Processes

If contamination is present anywhere inside the deposition chamber, it will interact with the plasma and cause electrical instabilities during the coating process. Hydrocarbon contamination from oil, for example, will significantly change plasma chemistry and result in aberration of the plasma voltage and current. In some cases, arcing can occur due to collapse of the plasma sheath, followed by high-voltage isolation breakdown within the fixture elements inside the process chamber. Consider an incandescent light bulb operated with a crack in the glass bulb. The bulb is cracked and atmospheric gases, mostly nitrogen and oxygen, have seeped into the bulb and around the filament. As the filament begins to light up and glow, it quickly oxidizes and fails due to reaction with the atmospheric gases. A similar event takes place inside the ion plater chamber with contaminants present in the plasma. The ions and electrons in the plasma find a path of least resistance to balance current and voltage, resulting in an arc within the gas similar to lightning in the sky during thunderstorms. The path of least resistance of the electrons and ions through the contaminate atoms and molecules precludes a steady and sustained current flow through the plasma sheath to the surface of the balls.

High-voltage arcing and insulator breakdown result in plasma and deposition interruptions during the process and have a significant effect on microstructure and film stress. The more severe of these two plasma interruptions is from voltage isolation breakdown and is critically linked to preventative maintenance schedules and chamber process history.

There are several types of plasma models available to track current flow with varying degree of complexity and numerical intensity. A fluids-based model will be used in this chapter which is based on the concepts of fluids mechanics such as, continuity, conservation of energy and momentum, and equations of state. The fluids-based plasma model is more useful concerning manufacturing processes involving large quantities of substrates coated at one time. In comparison, particle-based models have been used extensively by researchers studying atomic movement within the plasma field. For example, the so-called particle-method self-consistent plasma model has been used to study electron and ion collisions within the sheath region of the plasma, Lieberman and Lichtenberg (1994). These models are used with quantum mechanical models to give insight to particle motion within the plasma. For example, researchers such as Qiu et al. (2008) use a particle method and a 300 VDC magnetron process at 5 mTorr to calculate electron velocity in the range of 10^7 m/s and an ion–electron collision frequency of about 10^8/s. They also observed that sheath thickness decreases with increase of both magnetron voltage and gas pressure. Their model accounted for argon ionization, elastic collision between electrons and ions, and argon ion excitation as the primary factors that affect film deposition.

Other researchers have focused more on the effects of ion impact to the substrate surface, also known as ion irradiation in the plasma literature. For argon ions that are positively charged, a negatively charged substrate will enable ion acceleration towards the substrate to satisfy the charge equilibrium requirement. Specifically, the ion-plating process uses momentum and atomic collision to improve coating adhesion by implanting deposited atoms into the lattice structure of the substrate. The momentum of the positively charged ions is driven in part by process voltage and the degree of ionization of argon within the plasma. To enable use of fluids-based models for deposition modeling, quantum-mechanics models associated with collisions can be used to account for ion-implantation depth, atom sputter removal, and defect diffusion within the substrate to name a few.

After a substrate has been bombarded with argon ions, how does one qualitatively measure the effects of ion irradiation on film performance? If the film is deposited on a flexible substrate and is on the order of a micron thickness, the film substrate will bend due to surface stress allowing a detectable change in film shape. Surface profile techniques can be used to detect substrate curvature during and after ion bombardment. Then, one may calculate the required stress to get the observed curvature and a link between bombardment and stress may be established. For example, Chan and Chason (2008) observed an immediate compressive stress during ion irradiation that reached steady state after 500 s. The stress immediately switched to tensile when the sputtering voltage was turned off. With this technique, they studied ion-irradiation effects on thin copper films by measuring the stress-induced curvature in the film during and after ion bombardment. The highlight of their work was to show that point defects generated by the ion collision cascades in the film layer causes bending stress within the film. They calculated that ions penetrate about 10–20 nm deep for the implantation energy in their tests. Prior to this work, it was believed that trapped argon molecules were the largest contributor to film bending stress.

Co-deposition processes allow multiple material layers to be deposited within the same film and during the coating process. The ion-plating process uses sputter deposition to implant a Ni–Cu layer to the ball surface. Then silver is evaporated onto the Ni–Cu layer with ion-implantation assistance to improve film adhesion.

Total pressure within the chamber is used to control the amount of evaporation deposition of the silver. As the pressure is increased, the silver is more encumbered to reach the ball surface and therefore less silver is applied. For comparison, the researchers of Morley et al. (2008) used a co-sputter and evaporation technique to control the amount of iron and gallium atoms applied to their film. They used a dc magnetron sputter system to deposit the iron and a resistance-heater crucible system to evaporate the gallium. The availability of both the Fe and Ga atoms was controlled by magnetron power and chamber pressure. With this process, they could control film constituent concentration.

6.2 Postdeposition Fatigue Testing

Thin-film coatings that will be used in rolling element applications such as in high-voltage and high-vacuum devices should be tested under conditions that replicate in-service use. The testing regime should include cyclic stress loading, mechanical vibration, and operating temperature conditions. For example, if a thin film of silver is to be used for solid lubrication of ball bearings for a rotating anode X-ray tube, then a test apparatus that will match the dynamic loading, vacuum, and heat conditions should be used. The RCF test rig of Chap. 4 is well suited to replicate in-service conditions.

If hard coatings such as titanium nitride (TiN) or boron nitride (BN) are to be applied to the rolling contact surfaces to reduce friction, the ball and coating should be tested under similar dynamic and temperature conditions as in-service use. Too often coating systems are tested in a laboratory setting using established ASTM and ASM test protocols and pass these tests only to fail quickly when applied to in-service applications. The authors have extensive experience testing thin-film coatings applied to ball bearings used in rotating anode X-ray tubes. For example, two well-known hard coatings, TiC and TiN, were applied to the races and balls of a bearing system followed by coating of solid silver for lubrication. The system was tested in linear cyclical fatigue as well as indentation testing and passed for over one million cycles. All the scratch and hardness data gathered in the lab suggested that the coatings, both the hard coating and the silver lubrication coating, would survive in-service bearing test for a least one million cycles. The in-service test had very low Hertz contact stress, 1.2–3.5 MPa, and about 120 °C temperature operation. This should have been an easy test for the coating system based on our fatigue, scratch, and indentation testing. Yet, 20 min into the test, or about 210,000 stress cycles when rotating at 8,400 rpm, we observed a spike in vibration amplitude at a frequency consistent with the ball-pass frequency of the bearing. After 40 min, the bearing vibration had exceeded a threshold consistent with bearing ball failure. The test did not even reach half of the expected life before the onset of failure. Posttest autopsy of the bearing balls, including SEM of the remaining coating system, revealed total delamination of the coating from the ball and race contact surfaces. After careful reexamination of the coating test processes, we noticed that none of the previous tests accounted for high-frequency excitation that occurs within the in-service bearing system. It was the slight imperfection of the balls and races combined with the rotation speed of the bearing system that produced the vibration that failed the coatings. The scratch and adhesion tests that were gathered did not account for this dynamic behavior during rolling contact and such high speeds, 8,400 rpm.

Thin-film coatings used in any rolling element bearing application should be tested in rolling contact fatigue. If the in-service use of the coating-system requires oil-based lubrication at atmospheric pressure, then testing in rolling contact fatigue with oil should be done before installing the coatings to an in-service application. For example, Liston (1998) and others used an oil-based RCF test rig to quantify spall failure and coating adhesion properties of TiN and BN coatings applied to M50 steel rods. The test was very similar to configuration 4 of Table 4.1, except that silver was not used as a lubricant. Instead, an oil-drip system with recirculation was used with a rotational speed of 3,600 rpm. Most notable from this testing was that a superlattice of alternating layers of TiN and BN coatings was applied using an unbalanced magnetron sputtering system. The custom microstructure greatly improved RCF life of the test elements. The Hertz contact stresses tested were 3.4 and 5.2 GPa. A 2–10-fold improvement of L_{10} life was observed for coating thicknesses of 250–500 nm. They also identified an optimal periodicity of the TiN and BN layers that further improved the L_{10} life.

Scanning electron microscopy (SEM) and secondary electron imaging (SEI) were used to assess the TiN and BN coatings after RCF testing. Autopsy revealed that about 60 % of the coating had flaked off within the first 20 h of testing for both coatings. But the remaining coating in the form of islands on the rod contact area still contributed to the increased L_{10} life. It is important to note that the damaged and flaking hard coating was subsequently removed from the rod-ball contact area by flowing oil from the drip process. Circulating oil was beneficial for this test because the recirculating oil carries away coating debris, and then the oil may be filtered and sent back to the contact area. This would be acceptable for testing so long as the in-service use allowed for similar removal of coating debris. In contrast, testing in vacuum using solid film lubrication is a closed system and the debris from flaking coating remains within the contact area for the life of the test. The use of liquid and flowing lubrication is not valid for coatings used inside high-vacuum and high-voltage devices.

Coating residual stress and surface hardness contribute to increased RCF life. For example, Liu and Choi (2008) report good agreement between test results and their model in areas of high compressive residual stress. High compressive stress reduces wear in that the coating does not deform under loading. In situations where high compressive stress was applied to the contact surface, there was a corresponding increase in RCF life. For this reason, Felmetsger et al. (2009) have investigated using gas-pressure control during deposition to effect change in residual stress in the film. Similar to the experiments of Chan and Chason (2008), they used precise control of argon pressure during deposition to induce compressive stress within the film, as well as insure film thickness uniformity. Indeed, ion flux within the dc-plasma discharge close to the substrate surface is primarily regulated by argon gas pressure. This relationship is explored in detail in Chap. 7.

Finally, some researchers have used the finite element method (FEM) to model RCF testing behavior. Bouzakis et al. (1998) simulated the contact area of a ball bearing and race to understand the fatigue stress loading that occurs as the ball passes over the race contact area. They correlated a model to experimental data from RCF tests to predict fatigue life performance of chromium nitride (CrN) and titanium aluminum nitride (TiAlN) deposited on steel rods. In their FEM model, they developed a quasistatic simulation of the RCF test considering just the ball and rod Hertz contact stresses. The rolling contact was modeled by repetitively applying a loading step followed by a relaxation step to the rod model. The pressure profile of each loading step was verified experimentally. The application and release of each load step in succession accounted for one ball pass over the substrate rod material.

6.3 Plasma-Assisted Deposition

The ion-plating process uses momentum transfer between positively charged argon ions and the metallic atoms to be deposited onto the substrate surface (Mattox 1972). Argon plasma is created by applying a dc-voltage potential to the substrate in presence of argon gas within a pressure range of 1–100 mTorr. The ionization of argon gas results in a purplish-blue color, and typically the sheath region near the substrate surface has the highest intensity glow. In comparison, nitrogen ions produce a pinkish-red color near the substrate surface. These colors can be seen in the aurora borealis in the night sky near the poles. A thin region of extreme voltage potential known as the plasma sheath forms over the surface of the substrate, between the quasineutral region and ball surface. Charge balance between the ball and argon gas takes place within the plasma sheath (Lieberman and Lichtenberg 1994).

Argon ions that are produced in the quasineutral plasma region eventually migrate to the edge of the sheath region and are then accelerated within the plasma sheath to velocities approaching 300 m/s in the direction of the substrate. The ions do not reside stationary within the plasma sheath; rather they are accelerated through the sheath. For this reason the plasma sheath region is darker than the semi-neutral region and it is the so-called dark space within the plasma. Figure 6.1 illustrates the location of the plasma sheath relative to the surface of the ball. The plasma outside of the sheath region is at 10–30 VDC, the so-called quasineutral plasma region. The voltage drop within the sheath region is much more severe, from the semi-neutral region of 10–30 V down to −1,500 VDC. It is within this region that the argon ions are accelerated towards the ball surface. The argon ions alone are not sufficient to balance charge between the argon gas and ball surface. Electrons pass through the sheath in the opposite direction of the ions to help maintain charge balance. Electrons supplied from the ball surface are accelerated into through the plasma sheath and into the semi-neutral region. It is these electrons

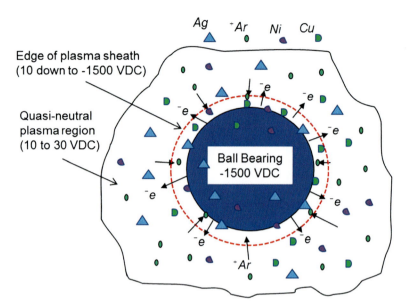

Fig. 6.1 Regions of the plasma near the surface of the ball

that cause intense argon ionization just outside the sheath region extending into the semi-neutral region. The fast electrons emanating from the ball surface ionize more argon gas and help to sustain the plasma glow on the outer edge of the plasma sheath.

Implantation of the deposition atoms occurs when the argon ions collide with deposition atoms, silver, for example, that are floating in the sheath region and are then accelerated to the substrate surface. The ionized argon atom smacks into the deposition atom and thereby imparts momentum to the silver atom, or other atom in its path, in the direction of the ball surface. The energy transferred from the argon ions to the coating atoms is sufficient to implant them into the lattice structure of the substrate material. In fact, the energy may be so severe as to damage and contaminate the film and substrate surfaces for bias voltages greater than 2 kV. We'll explore surface damage and ion mixing and their influence on RCF life in Chap. 7.

The energy of ion implantation is dependent on the relative mass of the argon ion and the metallic atoms that are to be implanted. The energy is related as

$$\frac{\Omega_t}{\Omega_i} = \frac{4\, m_i m_t}{(m_i + m_t)^2} \cos^2\theta, \tag{6.1}$$

where Ω_i and Ω_t are the incident and target kinetic energies of the colliding ions and atoms, θ is the angle of the collision, m_i and m_t are the incident and target mass, respectively. In this chapter the ion-plating process is explained in more detail with addition of numerical simulation of an ion-plating process using Simulink[TM]. But first data is collected for ion-plating processes using different pressures and voltages in order to validate the model. Figure 6.1 allows one to visualize what happens within the plasma and how voltage, pressure, and current affect deposition.

6.4 Coating Procedure for RCF Testings

The thin-film system and substrate used in this section are nickel–copper–silver deposited onto 7.94 mm ANSI T5 ball bearings using the dc ion-plating process. All ion-plating experiments were carried out in the system presented in Fig. 6.2. A process map of the baseline plating process recipe is presented in Fig. 6.3. Approximately 10 nm of nickel–copper is sputter-deposited onto the ball surface before silver is applied. Application of a nickel–copper layer to the ball surface is known to improve coating adhesion when working with steel-based ball bearings. After the nickel–copper layer, pure silver is evaporated and ion-plated to the nickel–copper surface on the ball. This process is a co-deposition process in that the nickel–copper is supplied by sputtering from a nickel–copper source, and the silver is supplied from evaporation and ion plating. Figure 6.4 illustrates the locations of the nickel–copper source and evaporating silver within the chamber. The nickel and copper atoms are sputtered off from the nickel–copper source plate above the carousel. The rate of supply of nickel and copper is controlled by the voltage applied to the source plate and carousel. Silver atoms rise from the bottom of the chamber after melting from the crucible. The availability of silver is controlled by chamber pressure and crucible heat.

The balls in this section were coated in the presence of pressure disturbances and arcing commensurate with those that will be simulated later in the chapter.

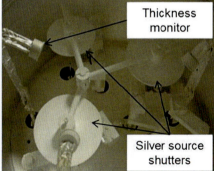

Fig. 6.2 Ion-plating system and view inside chamber (Reproduced with permission from J. Vac. Sci. Technol. A 29, 011005 Copyright 2011, American Vacuum Society)

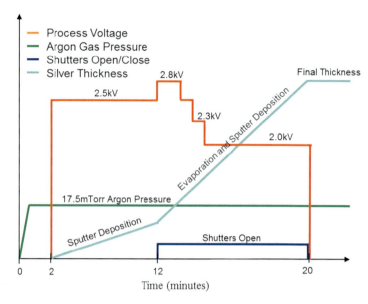

Fig. 6.3 Process map of baseline ion-plating process for DoE tests (Reproduced with permission from J. Vac. Sci. Technol. A 29, 011005 Copyright 2011, American Vacuum Society)

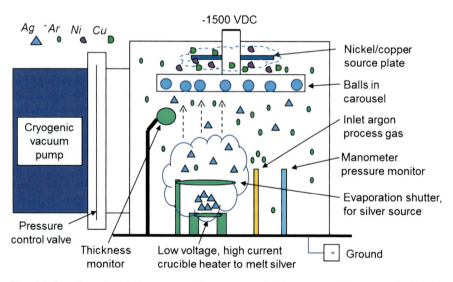

Fig. 6.4 Location of nickel–copper sputter source and silver evaporation source inside the chamber of Fig. 6.2

A sample of the balls from all coating runs were analyzed for coating thickness and layer composition. The rest were life tested in the RCF tester presented in Chap. 4. The rolling contact fatigue tests were carried out in high vacuum with a Hertz contact stress of 4.1 GPa at 130 Hz (7,800 rpm) rotational speed.

6.5 Plasma Effects on Coating Thickness

Voltage and pressure effects on film thickness were identified using regression analyses and central composite design with response surface analysis. A two-parameter, full factorial design-of-experiments (DoE) design was used over the variable factors of chamber pressure and process voltage. The deposition rate and thickness monitor parameters for all runs in the DoE were held constant. All DoE runs were stopped when the thickness monitor measured 110 nm. All DoE runs had a similar overall process time of about 20 min. The process voltage and pressure were adjusted as presented in Table 6.1. For example, experiment 3 was run at 17.5 mTorr pressure with the voltage profile in Fig. 6.3 shifted down by 1.0 kV. Experiment 12 was run at 15.0 mTorr chamber pressure with the voltage profile shifted up by 1.0 kV. Final thickness for each set of DoE balls was determined by weight comparison of pre- and post-coated balls described in Chap. 5. Sixty balls were used for each DoE set and an average thickness of nickel–copper–silver based on total weight was assumed for each ball. All balls for all the tests came from the same lot of ANSI T5 and were preprocessed the same way as the tests presented in Chap. 4.

The DoE coating tests were run with the dc-voltage supply in voltage-control mode using the system presented in Fig. 6.2. Operating the dc-voltage supply in voltage control allows for specific control of process voltage for varying pressure range. Since this is a dc-process, the total power input to the plasma is calculated as the product of current and voltage output.

Considering Table 6.1 and Fig. 6.5, at lower process voltages, the thickness monitor underpredicts true coating thickness. At higher voltages the thickness monitor overpredicts true thickness. These results are understood by considering the effect of process pressure on the motion of the evaporated silver from the crucible. From Fig. 6.4, the silver must migrate through the chamber up to the carousel and balls. At higher chamber pressures, there are more argon atoms and

Table 6.1 Run order and thickness results for DoE coating tests

Run order	kVolts	mTorr	Thickness nm
1	3.5	18.5	89.70
2	2.5	17.5	101.80
3	1.5	17.5	108.90
4	2.5	18.5	96.40
5	1.5	18.5	104.90
6	2.5	17.5	101.90
7	2.5	17.5	102.30
8	1.5	15.0	137.60
9	3.5	17.5	78.70
10	2.5	17.5	102.10
11	2.5	15.0	115.70
12	3.5	15.0	66.50
13	2.5	17.5	101.60

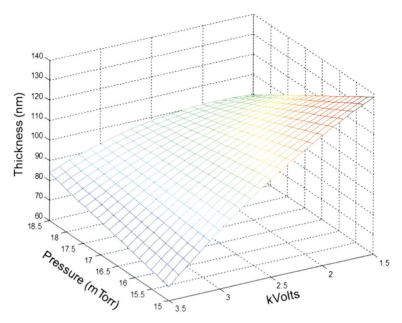

Fig. 6.5 Response surface plot of thickness DoE data of Table 6.1 (Reproduced with permission from J. Vac. Sci. Technol. A 29, 011005 Copyright 2011, American Vacuum Society)

ions present and the argon atoms simply get in the way of the migrating silver atoms. The result is less availability of silver to the plasma sheath for deposition. Increasing the process pressure will also reduce the plasma sheath thickness, which reduces the available volume for ions to impact the silver atoms onto the ball surface and further reduces coating thickness.

From Fig. 6.5 for voltages above 2.5 kV, the effects on coating thickness are dominated by the process voltage. This can be explained by the increased sputter energy above 2.5 kV that effectively removes some of the coating that has been deposited onto the balls. Higher process voltages increase the sputtering action within the sheath. The intense sputter action associated with process voltages above 2.5 kV accelerate the ions into the coating such that nickel, copper, and silver are removed from the ball surface, resulting in less final thickness.

6.6 Analysis of Extreme DoE Coating Tests

Table 6.2 contains elemental results of four runs that represent the extreme values of the DoE variables of Table 6.1. The nickel (Ni) and copper (Cu) present in the coating are from the Monel[TM] 400 sputter plate mounted above the substrate carousel as illustrated in Fig. 6.4. For comparison, elemental results using SEM for one uncoated ball is presented as well. The SEM, when used for surface

6.6 Analysis of Extreme DoE Coating Tests

Table 6.2 Coating composition using SEM of one ball for four DoE coating processes

Element	Baseline T5 Without coating	DoE Run 2 101.8 nm Coating peak Power 1,000 W 17.5 mTorr 2.5 kV	DoE Run 3 108.9 nm Coating peak Power 338 W 17.5 mTorr 1.5 kV	DoE Run 12 66.5 nm Coating peak Power 1,000 W 15.0 mTorr 3.5 kV	DoE Run 8 137.6 nm Coating peak Power 180 W 15.0 mTorr 1.5 kV
Ar	0.00	0.82	0.00	0.00	0.00
Ni	0.00	10.89	2.70	0.56	0.45
Cu	0.00	3.33	0.84	0.30	0.12
Ag	0.00	24.13	39.45	19.54	32.22
Fe	65.82	45.12	39.83	63.68	53.15
Cr	4.15	3.72	3.26	4.03	3.69
V	1.25	1.88	1.08	1.13	1.04
W	18.00	10.11	12.84	10.76	10.33
C	0.78	0.00	0.00	0.00	0.00
Co	10.00	0.00	0.00	0.00	0.00

Reproduced with permission from J. Vac. Sci. Technol. A 29, 011005 Copyright 2011, American Vacuum Society

analyses, is influenced by the substrate composition. A baseline analysis of the ball material near the surface was needed for comparison with the coated balls.

The Ni and Cu act as tracers to help determine the balance between sputtering and evaporation deposition as the process voltage and pressure are changed. For example, Run 2 has trace amounts of Ni and Cu in ratio similar to the composition the Ni–Cu source plate, suggesting that material from that source contributed to deposition thickness on the balls. The Ni and Cu source plate is above the carousel and once these atoms are sputtered off, gravity will assist with their migration into the plasma sheath and eventually implanted to the ball surface. Run 8, which used a lower-power and lower-voltage process than Run 2 has less Ni and Cu, suggesting that at 1.5 kV and 15 mTorr the sputter plate is not a significant source of deposition material. Lower voltage means less sputter deposition of Ni and Cu, and lower pressure means an easier path for the evaporated silver to migrate into the plasma sheath. In contrast, Run 12 was run at 3.5 kV and 15 mTorr and has a small amount of Ni and Cu present and less silver as well. This combination of voltage and pressure resulted in more severe sputtering in the sheath around the balls and gave a reduced overall coating thickness. Even for the same peak power as Run 2, the higher-voltage process of Run 12 resulted in an overall reduced thickness due to sputter removal from the ball surface.

The ratio of Ni to Cu in Run 3 is similar to that of Runs 2 and 8 with Run 8 having the least Ni and Cu overall. A comparison of the total amount of Ni and Cu among Runs 2, 3, and 8 suggests the influence of plasma power on sputter yield. Increasing plasma power may improve implantation but can also cause coating redistribution. For example, Run 8 has the least amount of Ni and Cu with plasma power of 180 W and Run 2 has the largest amount of Ni and Cu with a plasma power of 1,000 W. Kolev and Bogaerts (2009) have predicted an increase in ion energy as pressure decreases and also confirm that there is less redistribution of sputtered material with increasing gas pressure for the ranges of 4–100 mTorr. This suggests a voltage and pressure combination to simultaneously improve implantation while reducing sputter removal and redistribution. This voltage and pressure combination will be explored in Chap. 7. Finally, the presence of argon in Run 2 in Table 6.2 was confirmed during RCF testing in vacuum using residual gas analysis (RGA). The argon gas gets trapped inside the coating layers during deposition for process pressure of 17.5 mTorr and is liberated from the coating during the run-in portion of the RCF test.

6.7 RCF Testing of Extreme DoE Coated Balls

The rolling contact fatigue test platform presented in Chap. 4 was used to quantify the RCF life of the extreme DoE coated balls. The tests were conducted in high vacuum in the range of 10^{-7} Torr using a fixed load and speed for all tests. The Hertz contact stress was calculated as 4.1 GPa with a rotation speed of 130 Hz (7,800 rpm). For clarity, Fig. 6.6 contains a cross section of the cup and ball configuration and the proximity of the balls and rod for the RCF setup. This setup is the same as Fig. 4.1 in Chap. 4, except that the balls are coated with ion-plated nickel–copper–silver instead of evaporated pure silver.

In addition to the validation tests of Chap. 4, the RCF test platform was further validated for repeatability using multielement coatings on 7.94 mm balls for nine tests. These tests used a separate set of nickel–copper–silver-coated balls, all from the same lot and process history. The coated balls were purchased from an outside ball coating supplier and represent an outside coating system that is independent of the coatings presented in Chap. 4. A fatigue life curve of the repeatability tests is presented in Fig. 6.7. A screenshot of the data monitoring system results and a posttest photo of one ball is presented in the inset of Fig. 6.7 as well. The contact stress load was chosen using Fig. 4.12 so that all tests would end within 50 h running at 130 Hz rod rotation.

Five RCF tests from four DoE coating sets were RCF tested in high-vacuum conditions: Runs 2, 3, 8, and 12, as these represent the extremes of the DoE design space with respect to process pressure and voltage. Two categories of failure modes were observed when testing the DoE coatings in Tables 6.2 and 6.3. The first observed failure mode was nickel–copper–silver depletion followed by spallaing at the ball surface. These tests were considered successful in that the coating depleted instead of flaking off during the test. The average test times based on five RCF tests each were 13.3 h for Run 3 and 12.5 h for Run 8. The longest test time of Run 3 was 32.0 h and for Run 8 the longest time was 30.1 h. Surface spall due to depletion of nickel–copper–silver was observed in the repeatability tests in Fig. 6.7 as well. For example, the balls used in Fig. 6.7 consistently failed due to

6.7 RCF Testing of Extreme DoE Coated Balls

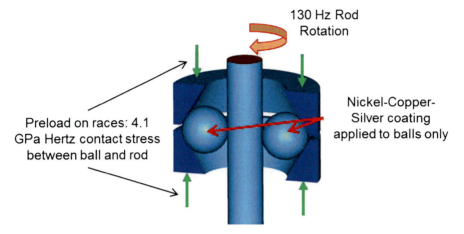

Fig. 6.6 Cross section of cup configuration with relative location of the ion plated Ni–Cu–Ag-coated balls

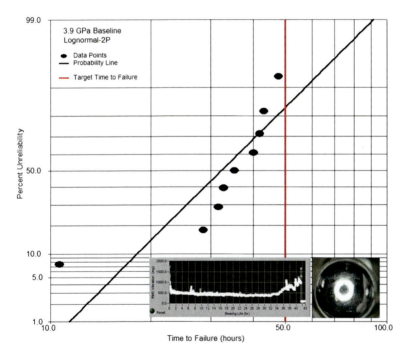

Fig. 6.7 Repeatability tests of the RCF tester using Ni–Cu–Ag-coated ANSI T5 balls purchased from Koyo™. Vibration history and one failed ball in lower right corner

nickel–copper–silver, or lubricant, depletion. The coating system did not fail. Rather the balls simply ran out of nickel–copper–silver and then spallaing occurred at the ball surface due to increased friction. This failure mode was

Table 6.3 RCF results of extreme DoE tests

		DoE Run 2	DoE Run 3	DoE Run 12	DoE Run 8
		101.8 nm	108.9 nm	66.5 nm	137.6 nm
		Coating peak	Coating peak	Coating peak	Coating peak
		Power 1,000 W	Power 338 W	Power 1,000 W	Power 180 W
		17.5 mTorr	17.5 mTorr	15.0 mTorr	15.0 mTorr
RCF test		2.5 kV	1.5 kV	3.5 kV	1.5 kV
1		8.1	8.2	3.3	8.0
2		6.5	32.0	4.5	7.1
3		6.9	8.5	3.1	30.1
4		6.6	10.1	3.7	5.5
5		6.4	7.5	4.8	11.9

Reproduced with permission from J. Vac. Sci. Technol. A 29, 011005 Copyright 2011, American Vacuum Society

observed in the test results of Chap. 4 as well, lubricant depletion followed by spalling at the ball surface.

The second observed failure mode was spallaing and failure of the coating within 3–8 h. This failure mode was unique in that the balls did not show evidence of precession during the test, that is, the balls did not roll over all surface of contact. The balls from Runs 2 and 12 exhibited the non-precession failure mode repeatedly. The non-precession balls did not run over the entire surface of the ball; instead, a single track was worn into the ball surface and failure followed within 8 h. This suggests a change in lubrication properties of the nickel–copper–silver coating for DoE Runs 2 and 12.

Balls from DoE Run 2 failed consistently around 6.9 h based on five RCF tests. The shortest test times of all tests came from Run 12 which had a failure in 3.1 h. These data suggest that increased process voltage decreases RCF life. More interesting was the nature of the failure of the coatings from Runs 2 and 12. These tests showed little ball precession suggesting different lubrication properties. Analysis and testing in Chap. 7 will address in detail the second observed failure mode. The processes used in DoE Runs 3 and 8 gave similar RCF results as the repeatability study and with Chap. 4 results in that the coating was depleted evenly over the entire surface of the ball.

6.8 Ion-Plating Model in Simulink™

Plasma sheath disturbances occur when arcing and subsequent pressure changes take place inside the process chamber. During these events, a poorly tuned controller or a poorly designed process can introduce pressure and voltage excursions in the process gas by either overshooting or undercorrecting for the event. In this section, a manufacturing ion-plating process is simulated using a matrix-sheath model and two types of process pressure control: proportional–integral–derivative (PID) and the linear quadratic regulator (LQR). The PID model represents a control scheme for a typical ion-plating process. The subsystem models were validated using the system shown in Fig. 6.2. Cathode sheath model validation consisted of comparison with published data from dc-cathode sheath calculations of Meyyappan and Kreskovsky (1990) and experiments of Fancey and Matthews (1990). Using the test data of the previous section for that range of process voltages and pressures, we can begin to build and fit a comprehensive process model to this experimental data.

6.8.1 Cathode dc-Sheath Model

At carousel potentials in the kV range, the cathode dark space of the plasma sheath is abnormal and the energy distribution of electron flux across the surface of the ball is assumed to be uniform. This example is illustrated in Fig. 6.1 using the dashed line at the edge of the plasma sheath. With this assumption, a 1-D sheath model will be used to calculate ion current density at the sheath and ball interface. For clarity, the surface of the ball is the cathode and the objective of this section is to calculate the current density at the ball surface. To be clear, there is an additional plasma sheath over the carousel as well, but the electron flux over the carousel surfaces is not uniform and is not needed for the present study. The carousel current density is constant and present for all tests, but does not adversely influence our study. A matrix-sheath model from Lieberman and Lichtenberg (1994) was chosen to simulate the plasma sheath parameters over the ball surfaces. The current density J at the sheath and ball interface on the balls is

$$J = e n_i \bar{u}_i, \tag{6.2}$$

where e and $\bar{u}_i$ are electron charge and ion group velocity, respectively, and n_i is the ion density within the sheath. The ion group velocity is calculated as

$$\bar{u}_i = \left(\frac{e V_0 \pi \lambda_i}{m_i s} \right)^{\frac{1}{2}}, \tag{6.3}$$

where V_0, λ_i, m_i, and s are the process dc voltage, ion mean-free path, ion mass, and sheath thickness, respectively. The ion mean-free path is affected by the total gas pressure in the chamber and is defined as

$$\lambda_i = \frac{1}{n_g} \sigma_i, \tag{6.4}$$

where n_g is the gas density and σ_i is the ion collision cross section. The parameter σ_i accounts for the quantum mechanics of the collisions within the sheath as found tabulated in the literature. The sheath thickness based on a 1-D approximation is calculated as

$$s = \sqrt{\frac{2\varepsilon_0 V_0}{en_i}}, \tag{6.5}$$

where ε_0 is the permittivity of free space constant and n_i is the ion density. The electric field E across the sheath is calculated as

$$E = \left[\frac{3en_i\bar{u}_i}{2\varepsilon_0}\right]^{\frac{2}{3}} \frac{x^{\frac{2}{3}}}{[2e\lambda_i/\pi m_i]^{\frac{1}{3}}}. \tag{6.6}$$

Equations 6.2, 6.3, 6.4, 6.5, and 6.6 are used in the ion-plating plant model in Fig. 6.8 to calculate ion current density and ion kinetic energy at the sheath and ball-surface boundary. A constant ion density n_i taken from a pre-sheath calculation is used in Eq. 6.5 to reduce numerical complexity.

6.8.2 Sputter and Evaporation-Sputter Deposition Models

As shown in Fig. 6.4, the process to be simulated has two mechanisms of deposition: (1) sputtering from a Monel 400 Ni–Cu source plate above the carousel and (2) evaporation and sputtering of silver atoms from a crucible below the carousel. The Ni–Cu source plate and carousel are connected and are always at the same potential. Evaporation-sputtering deposition begins when the source shutters are opened and silver atoms migrate into the plasma sheath, as shown in Figs. 6.1 and 6.4. Once inside the sheath, argon ions impact the coating atoms onto the ball surface. The shutters are situated over the silver crucible as shown in Fig. 6.2 and illustrated in Fig. 6.4. Deposition by sputtering begins as soon as the process voltage (kV) is turned on and the plasma is established as shown in Fig. 6.3. The first layers of Ni and Cu and subsequent Ag atoms are sputtered on and their properties are influenced by the plasma voltage and current.

Nickel and copper atoms are liberated from the Monel 400 source plate at a rate R_{sp} calculated as

$$R_{sp} = Y_{sp} V_{sp} \frac{J}{q}, \tag{6.7}$$

where Y_{sp}, V_{sp}, J, and q are the sputter yield, atomic volume, current density, and charge, respectively. Since the balls are at the same potential as the Monel 400 source plate, silver and coating may be sputtered off from the ball surface as well. Equation 6.7 may be used to account for sputter removal of coating from the ball surface for higher-kV processes. The only difference between sputtering Ni,

6.8 Ion-Plating Model in Simulink™

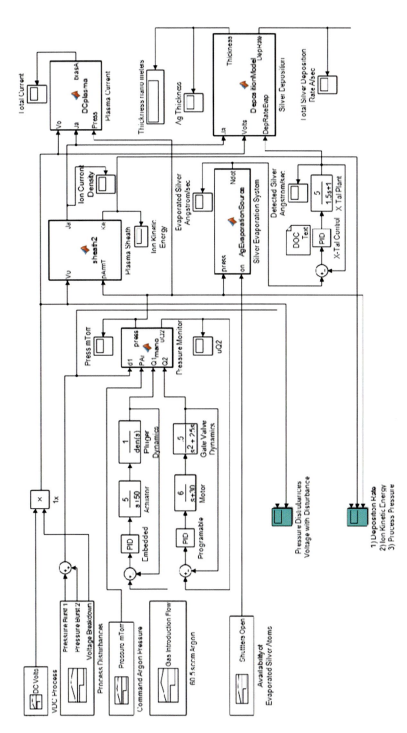

Fig. 6.8 Ion-plating process model in Simulink using PID control

Cu, and Ag from any surface is accounted for in the sputter yield Y_{sp} for each atom type. For example, for argon gas and silver, the sputter yield is about 0.7 for 100 eV ions. In the ion-plating simulation, Eq. 6.7 accounts for two sputter removal processes: sputter removal of nickel and copper from the Ni–Cu source plate and sputter removal of silver from the ball surface. The controller design parameter for Eq. 6.7 is the current density, J, which is influenced by the process voltage and argon gas pressure.

6.8.3 Elements of the Ion-Plating Simulation Model

The evaporated silver available inside the chamber is modeled using a Hertz–Knudsen model from Mattox (1998). The Hertz–Knudsen model uses the temperature and vapor pressure of silver to calculate the amount of silver leaving the crucible as function of time:

$$\frac{dN}{dt} = \frac{\varphi\sqrt{T}}{\sqrt{2\pi m}}(p^* - p(t)). \tag{6.8}$$

The output dN/dt has units of atoms/cm^2/s and with m, T, φ, and p^* defined as the mass, temperature, view-factor constant, and vapor pressure of the liquid silver, respectively.

The plant model and controller for the thickness monitor system is taken from the literature describing a typical system as found in a Sigma Instruments (2007) thin-film deposition controller manual. The monitoring system plant modeled is as

$$G(s) = \frac{K_p \exp(-L_d s)}{\tau s + 1}, \tag{6.9}$$

where K_p, L_d, and τ are the plant gain, system dead time, and time constant, respectively. Typically, the thickness monitor is placed as close as practicable to the substrates inside the chamber and some calibration and tuning are required for accurate measurements. If the process voltage or pressure changes, then the thickness monitor needs to be re-tuned for the new process parameters. During arcing and pressure burst events, the thickness monitor measurement can be very inaccurate. A robust PID controller is used in the model and is described as

$$C(s) = K_c\left(1 + \frac{1}{T_i s} + T_d s\right), \tag{6.10}$$

where K_c, T_i, T_d and C are the PID controller gain, integral time, derivative time, and transfer function, respectively. The process is modeled in Simulink^{TM} and is presented in Fig. 6.8.

6.8 Ion-Plating Model in Simulink™

The process pressure is maintained by two gas flow systems acting on the chamber simultaneously. A vacuum pump and conductance valve account for the gas removal portion of the pressure control system and a gas introduction valve is used to introduce argon into the chamber. The volume flow rate in this simulation for gas introduction was 100 sccm (standard cubic centimeters per minute). One common mass flow instrument uses a solenoid and field actuation coil to meter gas input, the details of which can be found in O'Hanlon (1989). The gas introduction system is presented below as a third-order state space system that includes solenoid dynamics and a field coil actuator model:

$$\{\dot{x}\} = \begin{bmatrix} 0 & 1 & 0 \\ \dfrac{-k}{m} & \dfrac{-c}{m} & \dfrac{K_{mx}}{m} \\ 0 & 0 & \dfrac{-R_l}{L} \end{bmatrix} \begin{Bmatrix} x_1 \\ x_2 \\ x_3 \end{Bmatrix} + \begin{Bmatrix} 0 \\ 0 \\ L^{-1} \end{Bmatrix} V_x(t), \quad Q_{in} = \begin{Bmatrix} K_f \\ 0 \\ 0 \end{Bmatrix}^T \{x\}. \quad (6.11)$$

R_l and L are the resistance and inductance of the solenoid actuation coil, respectively. The constants K_{mx} and K_f account for field current and gas flow calibration factors and are typically determined for a specific gas and flow range. For example, when purchasing a gas flow meter, one must specify the gas type and desired flow range. Typically these meters are calibrated using nitrogen gas, and when used with other gases, one simply applies a correction factor. The gas input to the system Q_{in} is in units of sccm. The state variables $\hat{x}$ used in the simulation are the solenoid position, velocity, and field coil current.

The gas removal model uses a feedback PID controlled scheme in which the pump conductance is modulated based on pressure monitor feedback information. The conductance valve is illustrated in Fig. 6.4, between the cryogenic vacuum pump and process chamber. A capacitance manometer is used to measure pressure electrically by monitoring the position of a diaphragm in response to pressure changes inside the chamber. The conductance valve model of the gas removal system consists of a dc motor drive with a gear reduction and valve slide to increase or decrease conductance. The valve mechanism is modeled as a linear system that accounts for valve dynamics and motor actuation. The output variable is the exiting mass flow, Q_{out}, and is proportional to the valve-motor actuator position. The state space model is

$$\{\dot{z}\} = \begin{bmatrix} \dfrac{-R_l}{L} & 0 & \dfrac{-K_{ez}}{L} \\ 0 & 0 & 1 \\ \dfrac{K_{mz}}{J_l} & 0 & \dfrac{-b}{J_l} \end{bmatrix} \begin{Bmatrix} z_1 \\ z_2 \\ z_3 \end{Bmatrix} + \begin{Bmatrix} L^{-1} \\ 0 \\ 0 \end{Bmatrix} V_z(t), \quad Q_{out} = \begin{Bmatrix} 0 \\ 1 \\ 0 \end{Bmatrix}^T \{z\}, \quad (6.12)$$

where K_{mz} and K_{ez} are the conductance valve plant constant and motor constant. The state variables $\hat{z}$ are motor current, valve angular position, and angular velocity, with J_l and b as the mass moment of inertia and velocity damping, respectively.

The total pressure inside the process chamber is proportional to the difference of the incoming and exiting mass flows divided by the maximum pumping speed of the cryogenic vacuum pump (O'Hanlon 1989). The pressure model is described as

$$P = \frac{Q_{in} - Q_{out}}{S}, \qquad (6.13)$$

where S is the throughput of the pumping system at process pressure. The comprehensive process model is presented in Fig. 6.8 using Simulink.

6.9 Process Model Simulation

Arc disturbances that result in pressure increases and plasma collapse can be accounted for in the ion-plating model simulation. Considering the results in Table 6.3, deposition at 2.5 kV and above has a detrimental effect on coating life when tested in rolling contact fatigue. Deposition above 17.5 mTorr pressure also reduces RCF life of the coating. Based on these results the pressure and flow controllers must mitigate disturbance effects during deposition. Disturbance events that result in pressure and voltage excursions were simulated on the plant model in Fig. 6.8 for two controller schemes applied to the pressure monitoring system. In a dc ion-plating process that is run in a constant power mode, pressure disturbances will result in voltage and current changes in the plasma as well. The plasma power controller will adjust the process voltage to maintain constant power during the pressure event. For this reason, two controllers are simulated and applied to the pressure control system: a proportional–integral–derivative (PID) and linear quadratic regulator (LQR).

The models presented in Eqs. 6.2, 6.3, 6.4, 6.5, 6.6, 6.7, 6.8, 6.9, 6.10, 6.11, 6.12, and 6.13 are useful for process development and equipment capability. For example, ionization rate can be calculated using plasma-current density at the sheath boundary, which is dependent on process voltage. An ionization rate of 4×10^{15} ions/s is needed to sustain an argon plasma at 60 mTorr. This ionization would require a current density of 0.3 mA/cm^2 at 2,000 VDC. A system design engineer would need to select a chamber and control hardware to deliver the needed current density and voltage to sustain the needed plasma. The process model presented in Fig. 6.8 has been validated to thickness only. Plasma properties and stresses and defects created within the film over the range of pressure and voltages in this chapter are explored in Chap. 7.

Thickness and deposition rate data from the system presented in Fig. 6.2 is shown in Figs. 6.9 and 6.10, for comparison with the model presented in Fig. 6.8. The process is slightly different from the DoE process map presented in Fig. 6.3 in that the deposition time after sputtering is longer and the resulting final thickness is higher. There is very good agreement between the model and experimental results in Fig. 6.9 for all times except during a portion of the sputter deposition

6.9 Process Model Simulation

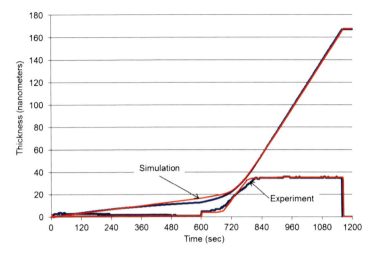

Fig. 6.9 Comparison of thickness and deposition rate for the system of Fig. 6.2 and the model in Fig. 6.8 (Reproduced with permission from J. Vac. Sci. Technol. A 29, 011005 Copyright 2011, American Vacuum Society)

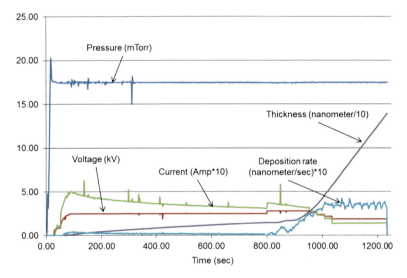

Fig. 6.10 Experimental data from the system in Fig. 6.2. The data has been scaled to fit on one graph

section of the process near 600 s. The model in Fig. 6.8 does not account for changes in the secondary electron emission of the ball surface and carousel that occurs during the first 600 s of the process. Secondary electron emission is illustrated in Fig. 6.1 as ^-e, the electrons leaving the ball surface to satisfy charge equilibrium in the plasma sheath. As a result, the slope of the deposition rate and the thickness deviate a little between 400 and 800 s of the process. Also, a constant sputter yield of 0.7 was used in the sputter deposition model and 0.9 was used in the sputter removal model on the carousel. During the actual process, however, these sputter yields change in relation to ion energy and secondary electron yield from the surface of the balls.

The decrease in plasma current over time in Fig. 6.10 is due to the removal of oxides on the carousel and ball surfaces. Each time the chamber is open to atmosphere, oxides will form on all surfaces, including the balls and carousel. The oxides have a higher secondary electron yield than the underlying surface of the balls and carousel, which results in a larger plasma current at the beginning of the process. As the process proceeds, the argon ions sputter off the oxides from the surface and the plasma current settles to levels associated with the secondary electron yield of the substrate carousel and ball materials. Process pressure, current, and voltage data presented in Fig. 6.10 contain arcing events and their influence on the plasma current and total pressure during a process. The model presented in Fig. 6.8 does not account for oxides on the substrate surfaces and the associated change in secondary electron yield. We will explore how to track ion density, ion and electron current, and electron density directly in Chap. 8.

An LQR control scheme following Burl (1999) was implemented on the gas removal system presented in Eq. 6.12 and used in the model presented in Fig. 6.8 in a 45 s simulation of the ion-plating process. The goal of the simulation was to quantify controller response to pressure and voltage disturbances, so the process time was reduced for study purposes. The model presented in Fig. 6.8 was used to calculate disturbance effects on plasma properties as well. The input disturbances include two pressure bursts, a pressure excursion, and one high-voltage breakdown event. The disturbances are presented in Fig. 6.11, and these are similar to the real disturbances observed in an actual process shown in Fig. 6.10. The output responses of the simulated ion-plating system for the two controller types are presented in Figs. 6.12, 6.13, and 6.14 for comparison.

The LQR control scheme allows performance optimization of the conductance valve by assigning penalties to the control law variables. For example, in the present study there is no way to completely mitigate pressure bursts inside the process chamber due to their rapid and random nature. However, system behavior after the burst subsides, usually within 1 or 2 s, may be controlled such that the control system need not further react to the event and cause a controller overshoot. The effects of the pressure burst on the coating process can be reduced with appropriate controller action. The difficulty lies with the inertia properties associated with the vacuum valve hardware. The LQR controller used in this study was optimized for minimum overshoot.

In contrast, the PID controller tries to respond to the pressure disturbance and subsequently overshoots the process pressure as shown in Fig. 6.12. The pressure disturbance then affects deposition rate and ion kinetic energy as shown in Figs. 6.13 and 6.14. Controller overshoot during deposition leads to the pressure and voltage combinations shown in Table 6.1 and ultimately to the extreme DoE results presented in Tables 6.2 and 6.3. A nonuniform coating would result from

6.9 Process Model Simulation

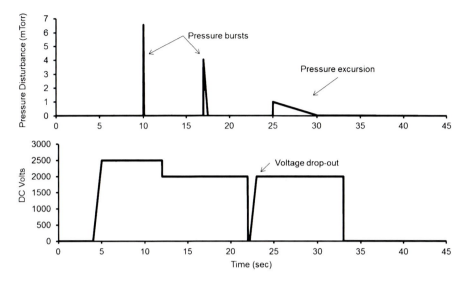

Fig. 6.11 Disturbance inputs and voltage profile to model in Fig. 6.8

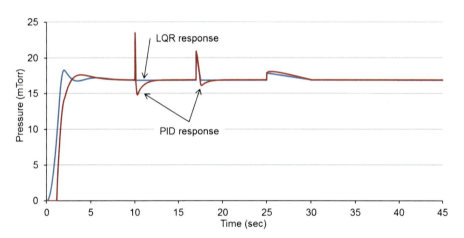

Fig. 6.12 Comparison of simulations using PID and LQR control of the conductance valve for model in Fig. 6.8

controller overshoot. Kolev and Bogaerts (2009) simulate a decrease in ion kinetic energy with increase in process pressure, due to the increased collisions with argon atoms associated with the pressure increase. No attempt was made here to correct for voltage disturbance; it is included in Figs. 6.13 and 6.14 to show its influence on the process output. Existing control technology of plasma power supply systems is sufficient to mitigate voltage disturbances.

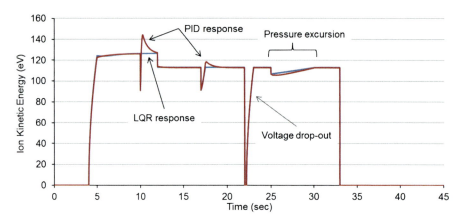

Fig. 6.13 Comparison of ion kinetic energy using PID and LQR control

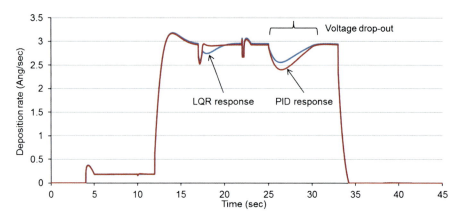

Fig. 6.14 Deposition rate response to the disturbances of Fig. 6.11 using PID and LQR control and the model of Fig. 6.8

Conclusions

For the process recipes studied in this chapter, the combination of 17.5 and 15 mTorr argon pressure with a plasma voltage of 1.5 kV yielded the longest average RCF test times, 13.3 and 12.5 h. The combination of 17.5 mTorr and 2.5 kV yielded an average RCF test life of 6.9 h and consistently exhibited a non-precession-type failure mode for all tests. These data suggest that higher process voltages during deposition have a detrimental effect on coating lubrication properties. A higher-density film is deposited at 2.5 and 3.5 kV, but it is unclear whether or not film stress, film density, or changes in the

(continued)

(continued)

coating composition have a detrimental effect on RCF life; these will be explored in Chap. 7. In contrast, a 1.5 kV process voltage had a positive influence on RCF life.

Ion-plating processes require multiple systems to function simultaneously during pressure burst and arcing events. Since the subsystem PID controllers are tuned around one specific operating point, errors can occur in the system response if chamber conditions change during the process. Plant disturbances such as pressure bursts due to arcing cannot be mitigated a priori and an optimization strategy like the LQR control could be used to reduce disturbance response impact.

References

Bouzakis KD, Vidakis N, Mitsi S. Fatigue prediction of thin hard coatings on the steel races of hybrid bearings used in high speed machine tool spindles. ASME J Tribol. 1998;120:835.
Burl J. Linear optimal control. Reading: Addison-Wesley; 1999.
Chan WL, Chason E. Stress evolution and defect diffusion in Cu during low energy ion irradiation: experiment and modeling. J Vac Sci Technol A. 2008;26:44–51.
Fancey KS, Matthews A. Evaporative ion plating: process mechanisms and optimization. IEEE Trans Plasma Sci. 1990;18:869–77.
Felmetsger VV, Laptev PN, Tanner SM. Innovative technique for tailoring intrinsic stress in reactively sputtered piezoelectric aluminum nitride films. J Vac Sci Technol A. 2009;27:417–22.
Kolev I, Bogaerts A. Numerical study of the sputtering in a dc magnetron. J Vac Sci Technol A. 2009;27:20–8.
Lieberman M, Lichtenberg A. Principles of plasma discharges and materials processing. 2nd ed. New York: Wiley; 1994.
Liston M. Rolling contact fatigue properties of TiN/NBN coatings on M-50 steel. In: ASTM, editor. ASTM STP 132. Ann Arbor: ASTM; 1998. p. 499–510.
Liu C, Choi Y. Rolling contact fatigue life model incorporating residual stress scatter. Int J Mech Sci. 2008;50:1572–7.
Mattox D. Structure modification by ion bombardment during deposition. J Vac Sci Technol. 1972;9:528–32.
Mattox D. Handbook of physical vapor deposition processing. Westwood: Noyes; 1998.
Meyyappan M, Kreskovsky JP. Glow discharge simulation through solutions to the moments of the Boltzmann transport equation. J Appl Phys. 1990;8:1506–12.
Morley NA, Yeh SL, Rigby S, Javed A, Gibbs MRJ. Development of a cosputter-evaporation chamber for Fe–Ga films. J Vac Sci Technol A. 2008;26:581–6.
O'Hanlon JO. User's guide to vacuum technology. 2nd ed. New York: Wiley; 1989.
Qiu Q, Li Q, Su J, Jiao Y, Finley J. Magnetic field improvement in end region of rectangular planar DC magnetron based on particle simulation. Plasma Sci Technol. 2008;10:694–700.
Sigma Instruments. SQC-300 thin film deposition controllers. Fort Collins: Sigma Instruments; 2007.

Chapter 7
Effects of Process Parameters on Film RCF Life

Abbreviations

$C_{Ar}(z,t)$	Argon concentration
D_{Ar}	Argon diffusion coefficient
Y_{Ar}	Argon entrapment defect yield
Υ_{Ar^+}	Argon-ion energy
M_a	Atomic mass of film material
k_b	Boltzmann constant
n_e	Electron density
T_e	Electron temperature
ρ	Film density
σ	Film-stress planar
a_p	Implantation depth
$C_I(z,t)$	Interstitial defect concentration
D_I	Interstitial diffusion coefficient
f_p	Ion flux
ν	Poisson ratio
P_{mTorr}	Pressure in milli-Torr
λ_p	Recombination distance for interstitial and vacancies
$C_V(z,t)$	Vacancy defect concentration
D_V	Vacancy diffusion coefficient
E	Young's modulus
AES	Auger electron spectroscopy
DoE	Design of experiments
EEDF	Electron energy distribution function
SEM	Scanning electron microscopy
SRIM	Stopping range of Ions in matter

The goal of this chapter is to identify the optimal range of pressure and voltage to maximize the RCF life of ion-plated nickel–copper–silver on ball bearings. Test data from Chap. 6 suggested that an optimal range exists such that coating depletion, and not surface spall is the desired failure mode for an optimal coating process recipe. The trade-offs associated with optimal process parameters include (i) improved coating adhesion using higher-voltage plasma, (ii) reduced film stress with lower-voltage plasma, and (iii) coating contamination from element redistribution. These trade-offs must be considered within the context of a large scale ion-plating process in which, say, 300 ball bearings are coated within a single process.

Equipment that is capable of coating large numbers of balls per process presents unique challenges for structural support and contamination risk. For example, the operational temperatures associated with the ion-plating process will dictate that all internal fixtures need to be strong and dimensionally stable up to 300 °C. Operation in high vacuum will require low-outgassing materials such as 304 stainless steel and OFHC (oxygen-free high-thermal conductivity) copper. However, the presence of stainless steel fixtures and OFHC copper in the process chamber exacerbates the contamination issue associated with high process voltages. From Chap. 6 we know that during the ion-plating process, the fixtures are sputtered as well creating a contamination risk associated with high process voltages. Elements such as nickel and chromium from the stainless steel fixtures may be sputtered off and redistributed onto the ball coating. In this chapter, we'll quantify the contamination risk and identify what process voltages should be avoided.

7.1 Plasma Diagnostic Tool

The first step is to associate process voltage with ion current density and electron energy in the plasma. With this link one may then monitor process voltage and pressure to avoid plasma properties that cause elemental contamination and lead to reduced RCF life. Ideally though, installation of a monitoring device capable of continuous plasma monitoring, a Langmuir probe, for example, would be the best possible way to prevent plasma damage to the coated balls. A Langmuir probe is used to measure properties of the plasma, such as electron density, electron temperature or energy, plasma potential, and ion density. Langmuir probe systems may be purchased from vacuum equipment supplier, and the software and electronics included with the probe system can also calculate ion temperature and the electron energy distribution function (EEDF). The ion density and EEDF are helpful parameters to understand the effects of ion bombardment on the coatings of Chap. 6 that resulted in reduced RCF life.

7.1.1 Langmuir Probe Experimental Procedure

The Langmuir probe is a wire of known thickness and material properties that is inserted into either the sheath or the quasineutral region of the plasma such that it electrically couples with the plasma. A steadily sweeping voltage is applied to the probe tip while the current flow is monitored. The instrument then measures current as a function of the voltage during the sweep, the so called I–V curve, and a current and voltage relationship is established. The sweeping voltage is typically -30 to 5 V for a single-probe system. The I–V curve is then used to determine plasma properties by calculating the rate of change of current and the rate-of-rate of change of current as a function of sweeping voltage. Think of this process as generating families of functions all from the same initial I–V data collection. The I–V curve and its derivatives are then fitted to a Taylor series expansion for current, given as

$$I(V+\delta V) = I(V) + \delta V \frac{dI}{dV} + \frac{1}{2}\delta V^2 \frac{d^2 I}{dV^2} + \ldots \tag{7.1}$$

Equation 7.1 may then be used to calculate plasma properties based on experimental data from the Langmuir probe. Figure 7.1 contains an I–V curve and its derivatives on the same plot for DoE Run 2 of Table 6.1. For more information concerning the current calculations, the plasma research literature has tutorials and many publications for calculation of plasma parameters using a Langmuir probe (Ruzic 1994).

The probe system used in this chapter to measure plasma properties is the single-electrode ALP-150 instrument from Impedans^{TM}. Figure 7.2 shows a probe tip inside the vacuum chamber and its proximity to the ball carousel fixture. The tip is approximately 20 mm from the surface of the balls, and the plasma data will be taken in the quasineutral region. The probe is used to measure properties of the plasma during the ion-plating process for each test condition in the DoE data of Table 6.1. Ion flux values for each of the DoE runs in Chap. 6 are presented in Table 7.1.

A word of caution concerning ion flux measurements is needed. The Langmuir probe systems will output ion and electron current density as a standard feature. It is up to the user to calculate ion flux for the specific fixture hardware and substrate. Since ion current flow to the ball surfaces is included in the ion flux measurement, the measurement and calculation are only valid for the specific number of balls present at the time of the measurement and for the specific fixture used at the time of the measurement. The ion flux will change as the number of balls changes per coating process. Any changes to the fixture(s): shape, material, or volume changes will require a recalculation of ion flux. With this in mind, when using the ion-plating process, it is best to maintain a consistent number of balls per coating run, in addition to specific requirements for process pressure and voltage.

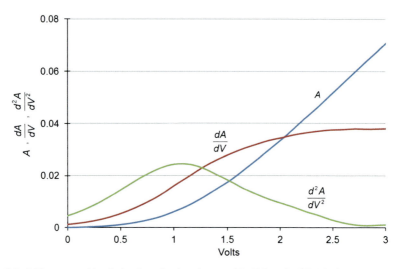

Fig. 7.1 I–V curve and its derivatives for the plasma of DoE Run 2 of Table 6.1

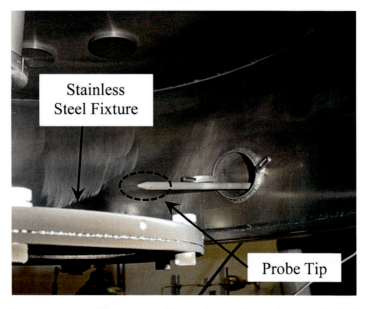

Fig. 7.2 Langmuir probe and its location relative to the stainless steel carousel inside the plasma chamber (Reproduced with permission from J. Vac. Sci. Technol. A 30, 031502 Copyright 2012, American Vacuum Society)

7.1 Plasma Diagnostic Tool

Table 7.1 Ion flux measurements from test conditions for DoE runs of Chap. 6

Run order	kVolts	mTorr	Argon-ion flux (cm^{-2} s^{-1})
1.	3.5	18.5	8.49E+15
2.	2.5	17.5	2.24E+15
3.	1.5	17.5	5.81E+14
4.	2.5	18.5	5.40E+15
5.	1.5	18.5	1.44E+15
6.	2.5	17.5	2.28E+15
7.	2.5	17.5	2.27E+15
8.	1.5	15.0	4.38E+13
9.	3.5	17.5	4.31E+15
10.	2.5	17.5	2.29E+15
11.	2.5	15.0	3.30E+14
12.	3.5	15.0	6.24E+14
13.	2.5	17.5	2.32E+15

Reproduced with permission from J. Vac. Sci. Technol. A 30, 031502 Copyright 2012, American Vacuum Society

7.1.2 Choosing an Appropriate Diagnostic Model

Before taking plasma diagnostic measurements, some preliminaries are needed to begin to understand the plasma. Low-temperature plasmas are initially understood by two characteristic lengths, the Debye length and the mean-free-path length. The Debye length is the maximum distance over which the ion will influence its neighbors. Beyond the Debye length, the ion's electrostatic effect does not impact neighboring particles. Low-voltage plasmas used in the ion-plating process have sheaths that are on order of a few Debye lengths thick. The mean-free path is the distance that the ion may travel before colliding with neighboring neutral atoms including argon, silver, copper, or nickel. The mean-free path is mostly influenced by total gas pressure during the coating process.

Plasma sheaths may be divided into four groups: (i) collisional-thin sheath, (ii) collisional-thick sheath, (iii) collisionless-thin sheath, and (iv) collisionless-thick sheath. All of these are influenced by pressure and voltage and may be selected based on calculation of Debye length and mean-free path. For the ion-plating process, the collisionless-thin sheath is the optimal choice. The collisionless-thin sheath allows for most of the ion's kinetic energy to be imparted to the neutral metallic atom for implantation to the ball surface. In contrast, the collisional-thick sheath is the worst condition for ion implantation. In this sheath condition the ion kinetic energy is consumed by numerous collisions before impacting the atoms to be deposited on the ball surface. For perspective, any one of these plasma conditions may exist during a coating process due to process aberration and controller disturbance.

Calculations of the Debye length and electron-neutral plasma lengths suggest a collisionless-thin sheath condition for all but DoE Run 8 which is calculated to be a collisionless-thick sheath. Following the example of Ruzic (1994), the electron-neutral mean-free path and the Debye length are calculated as

$$\lambda_o = \frac{0.061}{P_{mTorr}}, \qquad (7.2)$$

$$\lambda_D = 7,430\sqrt{\frac{T_e}{n_e}}, \qquad (7.3)$$

where T_e, n_e, and P_{mTorr} are the electron temperature (eV), electron density (m^{-3}), and argon gas pressure in mTorr, respectively. The probe tip radius used for all measurements was 0.195 mm. Processing in the collisionless-sheath regime allows more accurate calculation of plasma properties such as ion flux and electron temperature. In contrast, processes in the collisional-sheath regime are problematic in that one has no reliable method to account for electron-neutral collisions within the sheath. In the collisional-sheath regime, it has been confirmed that as process pressure increases, the sheath thickness becomes greater than the electron-neutral mean-free path and the presence of neutral gas atoms influences the sheath properties. The goal of these plasma tests and the ion flux data in Table 7.1 is to calculate a response surface regression for ion flux that will be used to calculate film stress as a function of process pressure and voltage.

7.1.3 Correct Interpretation of Langmuir Probe Data

Ion density is influenced by argon pressure and process voltage. There is a balance between the available amount of argon atoms to ionize and the charge balance to instigate electrons to enter the plasma from the ball surface. The plasma is sustained by electrons accelerating through the sheath and into the quasineutral region of the plasma. The electrons electrically couple with the ionized gas and so sustain the ionization process. Figure 7.3 contains a comparison of ion density using different combinations of argon pressure and voltage. Four pressures are considered over a voltage range of 1.0–2.5 kV.

There is a strong dependency on plasma voltage for pressures over 16.5 mTorr. In the pressure range of 16.5–18.0 mTorr, there is ample supply of argon to ionize and the ion density increases with voltage. There is maximum ion density, however, specifically for the combinations of 2.0 kV at 17.5 and 18.0 mTorr . For these conditions the increasing voltage does not increase ion density. There are many more collisions in the sheath and the ionization process is encumbered. Lowering the pressure to 16.5 mTorr allows for higher voltage and an increasing ion density. Operating in this process pressure approaches a collisionless-thick sheath, similar to Run 8 of Table 7.1. Below 16.0 mTorr there are less argon atoms to ionize and correspondingly there is a lower ion density.

7.1 Plasma Diagnostic Tool

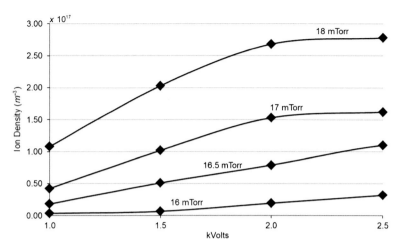

Fig. 7.3 Comparison of ion density for voltage and pressure

For comparison, plasma current and power are plotted in Fig. 7.4 as a function of voltage for the same process pressures presented in Fig. 7.3. The ion density of Fig. 7.3 tracks similarly to the plasma current, which is to be expected since plasma current is a driving factor for ion density. Plasma power however does not track with ion density; instead, as current increases the plasma power supply will simply increase the voltage to achieve the requested power. The comparison of Figs. 7.3 and 7.4 confirms that ion density does not track linearly with plasma voltage over a range of process pressure based on the measurements of two separate instruments: the Langmuir probe and the plasma power supply. What this means for the ion-plating process engineer is that there is a maximum achievable ion density based more on process pressure than on voltage. For example, it is very tempting to simply increase the process voltage in hopes to improve coating adhesion and ion implantation through increased plasma power. However, without increased ion density, the supply of ions to do the implantation, the plasma power increases and no benefit to adhesion is achieved. This was the condition observed for the DoE runs of Chap. 6 and Table 7.2 for process voltages greater than 2.5 kV. Increasing the plasma power via increased voltage did not improve adhesion nor RCF life.

The electron energy distribution function (EEDF) may be thought of as the fingerprint of the plasma. As the name implies, the EEDF is a probability distribution for the energy of the electrons that ionize the argon gas. This function is often modeled as a classical Maxwell–Boltzmann distribution, which is okay if the argon is pure and has very few contaminates. If gaseous molecules are present in the plasma in the form of contaminates, then the Maxwell–Boltzmann assumption is not valid and the data may represent a unique distribution other than Maxwell–Boltzmann. For example, low-temperature plasmas similar to ion plating are used for surface cleaning of surgical instruments. By nature of the process, the cleaning-plasma will contain contaminates and the EEDF may fit a Druyvesteyn function instead of Maxwell–Boltzmann. The EEDF is affected by voltage and pressure combinations, not just voltage alone. Figure 7.5 contains EEDF curves for four DoE

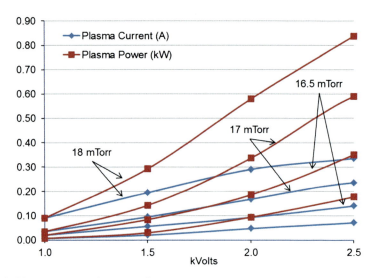

Fig. 7.4 Plasma current and power at four process pressures

Table 7.2 Comparison of ion flux and RCF life for DoE runs from Chap. 6

RCF test order	DoE Run 8, ion flux 4.38E + 13 (cm^{-2} s^{-1})	DoE Run 3, ion flux 5.81E + 14 (cm^{-2} s^{-1})	DoE Run 2, ion flux 2.28E + 15 (cm^{-2} s^{-1})	DoE Run 12, ion flux 6.24E + 14 (cm^{-2} s^{-1})
1.	8.0	8.2	8.1	3.3
2.	7.1	32.0	6.5	4.5
3.	30.1	8.5	6.9	3.1
4.	5.5	10.1	6.6	3.7
5.	11.9	7.5	6.4	4.8

Reproduced with permission from J. Vac. Sci. Technol. A 30, 031502 Copyright 2012, American Vacuum Society

runs from Table 7.1. The EEDFs of Runs 3 and 12 are similar in shape and energy distribution. Yet, the Run 12 combination was 3.5 kV at 15 mTorr and Run 3 was 1.5 kV at 17.5 mTorr. For this reason the, EEDF may be used to uniquely characterize low-temperature plasmas.

Results from Chap. 6 and Table 7.2 suggest that Run 3 gave the longest RCF life. Runs 12, 4, and 2 had early life failures and less RCF life than did Run 3. A comparison of the EEDFs for each DoE Run in Fig. 7.5 reveals that the EEDF of Run 3 is the more closely Maxwell–Boltzmann of the four EEDFs. The EEDF of Run 2 is wide and has an interesting bulge at about 2 eV. This suggests other processes were occurring over that energy range rather than just the ionization of argon. The EEDF of Run 4 has a bulge at about 1.5 eV, suggesting similar behavior as Run 2. Clearly, the Maxwell–Boltzmann-like distribution of Run 3 corresponds with improved RCF life, the pressure and voltage combination of 17.5 mTorr at 1.5 kV. Run 12 did not correspond to the optimal sheath condition; rather it approached the collisionless-thick condition and gave very poor RCF life results.

7.2 Plasma Effects on Film Properties and Composition 135

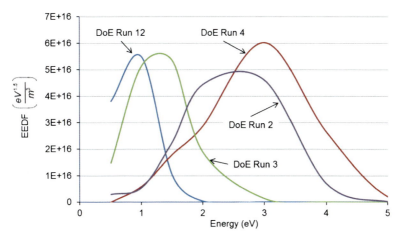

Fig. 7.5 Electron energy distributions for four DoE runs of Table 7.1

7.2 Plasma Effects on Film Properties and Composition

Plasma processes at voltages above 1.5 kV and within the pressure range of the tests presented in Table 7.1 suggest sputtering and implantation of argon gas atoms, as well as metal atoms from the stainless steel fixture used during the process. From Chap. 6, we found trace amounts of stainless steel components in deposition tests above 2.5 kV, suggesting that iron and nickel atoms from the steel fixture inside the chamber are being sputtered off and deposited along with the nickel–copper–silver coating. The increased presence of nickel and iron in the film correlated with reduced RCF life.

7.2.1 Ion and Elemental Mixing

To investigate the threshold of pressure and voltage for the onset of contamination from the fixture, several sputter deposition tests were carried out using Si_3N_4 balls to determine at what voltage and pressure contamination occurs. Si_3N_4 balls were used since they do not contain nickel, copper, or iron.

7.2.2 Auger Electron Spectroscopy Test

Figure 7.6 contains AES results from a process similar to DoE test 2 of Table 7.1 on a Si_3N_4 ball. The purpose of this test was to track and confirm that some of the nickel and all of the iron detected in the SEM results of Chap. 6 were not from the balls. Rather, the source of nickel and iron resulted from elemental sputter removal

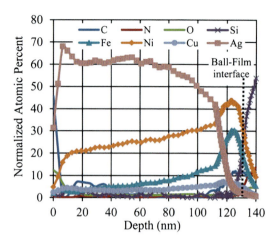

Fig. 7.6 AES depth profile of a 7.94 mm diameter Si$_3$N$_4$ ball that was coated with nickel copper silver using the process parameters of DoE Run 2 presented in Table 7.1 (Reproduced with permission from J. Vac. Sci. Technol. A 30, 031502 Copyright 2012, American Vacuum Society)

of these metallic atoms from the stainless steel fixture inside the chamber. The iron could only have come from the stainless steel fixture inside the chamber, while some of the nickel came from the Ni–Cu sputter plate along with copper. The process voltage at which the iron and nickel contamination would occur needed to be determined, and then we can compare that condition to the ion energy results generated using a simulation tool such as SRIM™ (Stopping Range of Ions in Matter) from Ziegler et al. (2008). These data are helpful in determining the voltage and pressure combinations that should be avoided to prevent film contamination whether by design or due to process aberration. To be clear, ion implantation and mixing near the ball surface is desirable to improve adhesion, but the results of Chap. 6 suggest that too much mixing has a detrimental effect on RCF life.

A comparison of ion flux with RCF life is presented in Table 7.2. It is clear that ion flux above 10^{15} (cm^{-2} s^{-1}) range reduces RCF test life. A closer examination of the data in Table 7.2 suggests that process voltage also influences RCF life. For example, the data from DoE Run 12 has a similar ion flux as Run 3, yet Run 12 has a lower RCF life suggesting an additional mechanism other than ion flux and internal stress is reducing life. Results from DoE Run 12 suggest that ion mixing plays a substantial role in reducing RCF life. DoE Run 12 used a process voltage of 3.5 kV, which is over twice that of Run 3 and is well over the 0.7 kV range used by Chan and Chason (2008) in their investigation of film stress and ion bombardment.

7.2.3 Ion and Element Implantation Modeling

SRIM results for process voltage of 2.5 kV predict implantation and mixing as much as 5–10 nm deep. Iron and the other constituents from the stainless steel fixture used to support the balls inside the chamber will be present throughout the

7.3 Film Properties Calculation

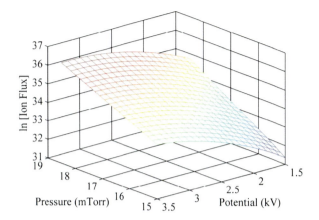

Fig. 7.7 Response surface of the natural log of ion flux in units of (cm^{-2} s^{-1}) from DoE in Table 7.1 (Reproduced with permission from J. Vac. Sci. Technol. A 30, 031502 Copyright 2012, American Vacuum Society)

coating for process voltages above 2.5 kV. It is known from Chan and Chason (2008) that ion flux in the range of 10^{14} cm^{-2} s^{-1} can cause stress in the film due to argon-ion implantation and defect creation. The RCF tests of DoE Run 12 are substantially less than the other three DoE conditions tested in Table 7.2 suggesting that RCF life was reduced not only from internal stress but that elemental contamination caused by ion mixing was a root cause of reduced RCF life as well.

Figure 7.7 contains a surface interpolation of the argon-ion flux from Table 7.1 as a function of process pressure and voltage. As argon-ion flux increases, some of the deposited film is sputtered off at voltages greater than 1.5 kV and at pressures within the DoE test conditions used in Table 7.1. The detrimental effects of increased ion flux on coating thickness were confirmed in the thickness results of Chap. 6. There was lower coating thickness as process voltage was increased, specifically for voltage greater than 1.5 kV. The Auger results of Fig. 7.6 and the reduced RCF life of DoE Run 2 in Table 7.2 suggest elemental layer mixing had a detrimental effect on RCF life.

7.3 Film Properties Calculation

The stress calculation in the film requires ion flux and implantation depth as a function of argon-ion energy. The SRIM tool is used to calculate the implantation depth and for the prediction of elemental mixing using the process parameters of Table 7.1. For clarity, SRIM uses two approximations for calculated results: (i) an analytic formula to determine atom–atom collisions and (ii) the free-flight path between collisions as presented in Ziegler et al. (2008). A center-of-mass coordinate system is used to reduce computational complexity and to limit the calculation to the collision of an incoming ion and a stationary substrate atom. A single atom model in an interatomic potential acting on the center of mass of the ion is used to calculate collision results.

Fig. 7.8 Argon-ion penetration depth model into the silver layer at 2.5 kV potential using SRIM™

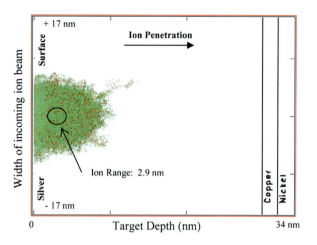

Fig. 7.9 SRIM™ ion penetration depth model of an ion-plating process similar to Run 2 of Table 7.1

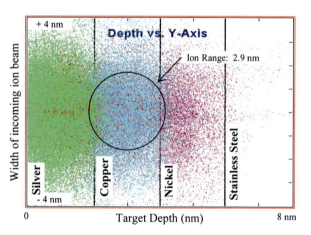

7.3.1 Layered Film Structure Properties

Thermal spike due to ion bombardment as reported by Kim et al. (1988) is not accounted for in the stress calculations in this section, nor in SRIM. Due to their atomic structure, copper and silver are considered good thermal spike materials within the voltage range of DoE runs of Table 7.1 due to their atomic structure, as reported in Norlund et al. (1998). Figure 7.8 presents results of one simulation at 2.5 kV using SRIM to determine ion implantation characteristics. A plot of ion implantation depth is presented in Fig. 7.9 from which the distribution-mean range and width may be extracted. The relative layer thicknesses are 30–2–2 nm of silver–copper–nickel. The steel ball surface is not shown since the ions do not penetrate beyond 15 nm at 2.5 kV. These data will be used for the film-stress calculations in the next section.

At ion flux greater than 10^{15} cm^{-2} s^{-1}, the SRIM results of Fig. 7.9 suggest elemental mixing of the layers based on elastic atomic collisions only. The AES

7.3 Film Properties Calculation

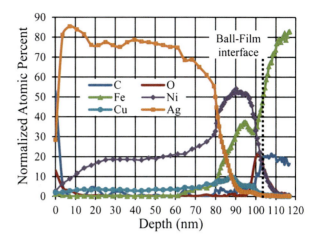

Fig. 7.10 AES depth profile from one ball of DoE Run 2 of Table 7.1 (Reproduced with permission from J. Vac. Sci. Technol. A 30, 031502 Copyright 2012, American Vacuum Society)

Table 7.3 Implantation data extracted from SRIM™ calculation

Process voltage, V	Implantation depth, a_p (nm)	Distribution width, w (nm)
1,500	2.2	1.3
2,500	3.0	1.8
3,500	4.0	2.3

results of Fig. 7.10 agree with the simulation results of Fig. 7.9 based on one ball out of 60 from DoE test Run 2 in Table 7.1. There is a mixture of nickel, copper, and silver at the interface due to ion mixing. For DoE Run 2, there is 10 min of nickel–copper sputter deposition at 2.5 kV at 17.5 mTorr pressure before silver atoms are introduced into the deposition process, following the process recipe of Fig. 6.3. The resulting thickness increase during the nickel–copper sputtering step was about 20 nm and consisted of nickel and copper with some iron present from the stainless steel fixture supporting the balls. The presence of nickel and iron from the stainless steel fixture is confirmed in Fig. 7.10 to be located between 80 and 100 nm, which correlates to the time in the process that voltage and pressure are held at 2.5 kV and 17.5 mTorr, respectively.

7.3.2 Film-Stress Calculation

The film-stress calculation in this section follows the work of Chan and Chason (2008) and is reproduced here for clarity. Ion flux and the kinetic properties needed for the calculations are presented in Tables 7.1, 7.3, and 7.4. The SRIM calculation used in the previous section to calculate implantation depth and distribution assumed a normal distribution. Table 7.4 contains kinetic properties published

Table 7.4 Kinetic parameters used in the film peak stress calculation (Chan and Chason 2008)

Defect type	Diffusion coefficient D (nm² s⁻¹)	Relaxation volume V (nm³)	Defect yield Y_{Ar} Y_I Y_V
Argon	3E-3	2.8E-3	1
Interstitial	3E-3	5.9E-3	5
Vacancy	6E-6	−5.9E-3	5

and calculated in the thin-film science literature, such as Chan and Chason (2008) and Ohring (2001).

To begin the stress calculation, a plane stress condition in the film is assumed and is given as

$$\sigma(z,t) = -\frac{1}{3}\frac{E}{1-\nu}(C_{Ar}(z,t)V_{Ar} + C_I(z,t)V_I + C_V(z,t)V_V), \quad (7.4)$$

where the parameters $C_{Ar}(z,t)$, $C_I(z,t)$, and $C_V(z,t)$ account for argon, interstitial, and vacancy concentrations, respectively. The defect concentrations are defined as

$$\frac{\partial C_{Ar}(z,t)}{\partial t} = \frac{f_p Y_{Ar}}{a\sqrt{2\pi}}\exp\left[-\frac{(z-a_p)^2}{2w^2}\right] + D_{Ar}\frac{\partial^2 C_{Ar}}{\partial z^2} - \frac{D_{Ar}C_{Ar}V_{Ar}}{k_B T}\frac{\partial^2 \sigma}{\partial z^2}, \quad (7.5)$$

$$\frac{\partial C_I(z,t)}{\partial t} = \frac{f_p Y_I}{a\sqrt{2\pi}}\exp\left[-\frac{(z-a_p)^2}{2w^2}\right] + D_I\frac{\partial^2 C_I}{\partial z^2} - \frac{D_I C_I V_I}{k_B T}\frac{\partial^2 \sigma}{\partial z^2}$$
$$- 4\pi\lambda_p C_I C_V (D_I + D_V), \quad (7.6)$$

$$\frac{\partial C_V(z,t)}{\partial t} = \frac{f_p Y_V}{a\sqrt{2\pi}}\exp\left[-\frac{(z-a_p)^2}{2w^2}\right] + D_V\frac{\partial^2 C_V}{\partial z^2} - \frac{D_V C_V V_V}{k_B T}\frac{\partial^2 \sigma}{\partial z^2}$$
$$- 4\pi\lambda_p C_I C_V (D_I + D_V). \quad (7.7)$$

The values for implantation depth a_p and distribution width w along with the concentrations and diffusion coefficients are tabulated in Tables 7.3 and 7.4. The ion flux f_p for each of the process runs is tabulated in column 4 of Table 7.1. The recombination distance λ_p between vacancies and interstitial defects was taken as 0.29 nm. Equations 7.4, 7.5, 7.6, and 7.7 are solved numerically using a PDE solver in Matlab. The code was validated by reproducing the results of Chan and Chason (2008) for defect concentration and stress to within 5 %. The code was then used to calculate the film-stress evolution for each of the DoE tests in Table 7.1.

The evolution of argon-ion concentration $C_{Ar}(z,t)$ for DoE Run 3 is presented in Fig. 7.11, namely, 1.5 kV at 17.5 mTorr pressure corresponding to an ion flux of 5.81×10^{14} cm⁻² s⁻¹ from Table 7.1. The ion depth is about 2 nm for ions impacting the silver layer for the 1.5 kV process. Response of the argon-ion concentration to beam-on and beam-off conditions can be seen on the left side of

7.3 Film Properties Calculation

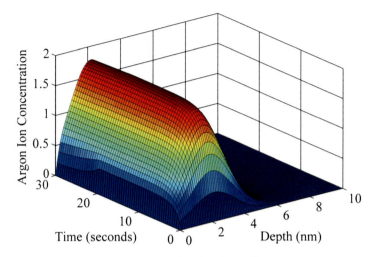

Fig. 7.11 Argon-ion concentration C_{Ar} (z,t), in units of (nm^{-3}) as a function of time and depth into the silver layer (Reproduced with permission from J. Vac. Sci. Technol. A 30, 031502 Copyright 2012, American Vacuum Society)

the surface plot in Fig. 7.11. When the beam is switched off after 20 s, the ion flux goes to zero and the argon begins to migrate within the silver layer and the concentration peak starts to decrease. This is consistent with the results of Chan and Chason (2008) for copper that were simulated as part of the code validation. The presence of argon migration in the silver layer was confirmed using residual gas analysis (RGA) data that was collected during the first 15 min of the RCF tests in Chap. 6. A decade-magnitude increase in pressure during the first 15 min of the RCF tests was observed due to liberation of argon gas from within the coating at high RCF stress levels. Simulation results for interstitial $C_I(z,t)$ and vacancy $C_V(z,t)$ concentrations were calculated as well as part of the full solution of Eqs. 7.4, 7.5, 7.6, and 7.7.

Peak stress during ion irradiation for each of the 13 DoE runs was calculated using Eqs. 7.4, 7.5, 7.6, and 7.7 and a corresponding surface regression fit was generated as shown in Fig. 7.12. The stress results compare inversely with the RCF life results of Table 7.2 such that peak compressive stress during deposition resulted in reduced RCF life. RCF life results shown in Table 7.2 suggest longer average life using a process voltage of 1.5 kV and reduced RCF life using process voltages of 2.5 kV and greater. From Table 7.1 the process voltage range of 1.5–2.5 kV corresponds to an ion flux in the range 10^{13} to 10^{15} cm^{-2} s^{-1}. These results exemplify the need for voltage and pressure control during deposition. Voltage or pressure excursions lasting even a few seconds can increase the ion flux to over 10^{15} cm^{-2} s^{-1} and have a negative impact on RCF life. Process pressure control will be discussed in Chaps. 8 and 9.

Internal film stress may also be calculated following the work of Ohring (2001) which considers argon-ion entrapment to be the primary cause of internal film

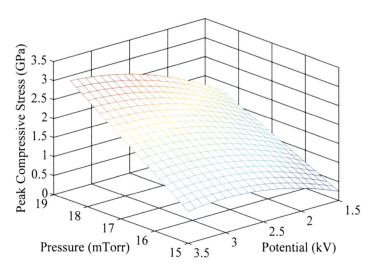

Fig. 7.12 Response surface of the peak compressive stress within the film using Eq. 7.4 and DoE data presented in Table 7.1 (Reproduced with permission from J. Vac. Sci. Technol. A 30, 031502 Copyright 2012, American Vacuum Society)

stress. Any thin film deposited with plasma influence and ion implantation in the thickness range of 10–100 nm will contain internal stresses in magnitude beyond the yield stress of the bulk material, usually on the order of 1–2 GPa. Low melting temperature metals such as silver, gold, copper, or lead will relax internal stresses quicker than slightly higher melting temperature metals such as nickel and iron. Internal stress relaxation has been attributed to thermal spike melting as reported by Kim et al. (1988), Norlund et al. (1998), and Mayr and Averback (2003), and in some cases to high- and low-mobility Volmer–Weber grain growth as reported in Ohring (2001). For thin-film application to thick substrates, such as a thin-film solid lubricant on the surface of a ball bearing, the substrate material properties are less important to internal film stress.

For quick calculation accounting for entrapped argon only, the film stress may be approximated as

$$\sigma = k \frac{f_p \sqrt{\Upsilon_{Ar^+} M_a E}}{(1-\nu)\rho}, \tag{7.8}$$

where f_p and Υ_{Ar^+} are the argon-ion flux and energy during processing, respectively (Ohring 2001). The material properties, M_a, ν, and ρ, are the atomic mass, Poisson's ratio, and the density of the film material. The parameter k is a scaling constant related to the test setup. Note that Eq. 7.8 considers argon-ion flux and ion energy only, discounting the effects of internal defects such as interstitials and vacancies and their recombination as accounted for by Chan and Chason (2008). Calculation of film stress using Eq. 7.8 and the ion flux from Table 7.1 is presented in Fig. 7.13

7.3 Film Properties Calculation

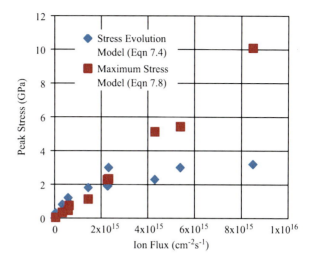

Fig. 7.13 Comparison of stress calculated from Eqs. 7.4 and 7.8 (Reproduced with permission from J. Vac. Sci. Technol. A 30, 031502 Copyright 2012, American Vacuum Society)

with comparison to the solution of Eqs. 7.4, 7.5, 7.6, and 7.7. Equation 7.8 shows a similar trending to that calculated from the stress evolution model of Eq. 7.4 up to an ion flux of 4×10^{15} cm^{-2} s^{-1}, but then begins to overpredict film stress. It is unlikely that the film could sustain stress above 3 GPa as is predicted for ion flux greater than 4×10^{15} cm^{-2} s^{-1} using Eq. 7.8. The stress evolution model of Eq. 7.4 is a better predictor of film stress because it accounts for defect recombination and argon migration. Migration and recombination reduce stress within the film.

Conclusions

The data and experiments in this chapter suggest an optimum range for process parameters during deposition to achieve maximum RCF life using the ion-plating process. RCF life has been shown to be an indicator of film quality with respect to internal stresses and elemental contamination within the context of large scale coating processes. The introduction of plasma processing has been shown to improve coating adhesion based on the data presented in Table 7.2 and has been suggested by Mattox (1998) and Ohring (2001) as well.

Evaporation of silver on to the balls as in Chap. 4 represents the purest silver lubricant coating possible, that is, no argon implantation, interstitial, or vacancy defects present due from the plasma. However, early spall failure may result due to poor adhesion and the lack of nickel–copper–silver implantation to the ball surface. At process voltages of 2.5 kV and greater, the stress present in the film is due to interstitial and vacancy defects, along with entrapped argon atoms from the deposition process. Thermal spike mixing occurs in the film layers for ion irradiation above 1.5 kV and has been shown

(continued)

(continued)

to reduce internal stress due to local melting. The thermal spike condition improves atom mobility and results in recrystallization and formation of the film structure. A trade-off emerges related to (i) the presence and implantation of the nickel–copper layer to improve adhesion and (ii) the lubrication aspects of a contamination-free silver layer. Deposition above 1.5 kV enables implantation with thermal spike to reduce film stress in nickel and copper resulting in improved surface adhesion. However, ion plating at 2.5 kV or higher risks contamination of the silver layer due to ion mixing from atoms in ball carousel fixture. These data suggest that the process voltage should not exceed 1.5 kV and that the ion flux should be maintained between 10^{13} and 10^{15} cm^{-2} s^{-1} for all combinations of process voltage and pressure.

References

Chan WL, Chason E. Stress evolution and defect diffusion in Cu during low energy ion irradiation: experiment and modeling. J Vac Sci Technol A. 2008;26:44–51.
Kim S, Nicolet M, Averback RS, Peak D. Low-temperature ion-beam mixing in metals. Phys Rev B. 1988;37(1):38–49.
Mattox D. Handbook of physical vapor deposition processing. Westwood: Noyes; 1998.
Mayr SG, Averback RS. Effect of ion bombardment on stress in thin metal films. Phys Rev B. 2003;68(214105):1–10.
Norlund K, Ghaly M, Averback RS, Caturla M, Diaz de la Rubia T, Tarus J. Defect production in collision cascades in elemental semiconductors and fcc metals. Phys Rev B. 1998;57(13):7556–70.
Ohring M. Material science of thin films, deposition and structure. 2nd ed. San Diego: Academic; 2001.
Ruzic D. Electric probes for low temperature plasmas. New York: AVS Press; 1994.
Ziegler J, Biersack J, Zieglar M. SRIM the stopping and range of ions in matter. Chester: SRIM; 2008.

Part III
Control and Disturbance-Rejection of Thin Film Deposition Systems

Chapter 8
Real-Time Process Control

Hardware-in-the-loop (HIL) testing of the ion-plating system in Chaps. 6 and 7 is presented in this chapter. Testing process control algorithms using system hardware enables faster process development. For example, a candidate controller model may be applied to a component subsystem, such as a vacuum control valve, or a mass flow control system that will be used during the deposition process. Simulation of the deposition process using real hardware rather than prototype components is possible using a suitable software tool such as Simulink or LabView, to name a few. All of the subsystem control algorithms within the coating process may be exercised using real hardware and validated during the testing and development phase of process development. After validation, the final source code may be deployed onto a correctly sized personal computer or similar firmware application.

In this chapter a real-time target application is used to test and characterize control schemes related to argon pressure control, plasma total-current monitoring, and argon process gas regulation to maintain optimum plasma properties during deposition. The primary benefit of HIL testing is controller development and optimization using real system hardware instead of models or prototypes. In addition, a second benefit is safe operation of the equipment components. The HIL process for controller development allows safe limits to apply to the hardware motion to prevent damage due to incorrectly tuned equipment.

8.1 Experimental Setup Using Simulink Real Time

All HIL testing in this chapter was carried out using the ion-plating system presented in Fig. 4.1 in Chap. 4. The ion-plating process requires multiple control systems to operate independent of each other at the subsystem level, but concurrently as a process. The Simulink Real-Time tool requires an external computing platform, the xPC host machine, from which the model will be executed. The xPC host allows interface with all component hardware through either analog or digital

I/O communication. One advantage of the host pc system within HIL testing is that any Windows-based pc can be made into an xPC host. Within the Simulink Real Time (SRT) method, the control model and diagram of the process are modeled in Simulink and then exported as executable code to the host pc. The code required to control the coating processes of Chaps. 4 and 6 can be loaded onto any Windows-based pc, allowing for performance evaluation of the host pc as well as the control algorithm itself.

A diagram of the specific components tested in this chapter is presented in Fig. 8.1. The ion-plating system requires one mass-flow instrument to inlet argon gas, Q_1. In addition, a second mass-flow instrument has been added to enable introduction of pressure disturbances, Q_d, to the chamber to exercise the control algorithm. The mass-flow controllers are used in analog control mode, meaning that they will accept 0–15 V dc input that is proportional to the requested gas flow in units of sccm. Chamber pressure is tracked using a capacitance manometer, such as the MKS 670A with signal conditioning. Exit flow conductance is controlled using a conductance valve in combination with a cryogenic vacuum pumping system. For comparison, the mass flow inlet Q_1 and the exit flow Q_2 were modeled as linear systems in Eqs. 6.11 and 6.12 in Chap. 6. The process plasma is generated using a dc power supply that has remote power-control capability. The plasma power supply has analog I/O that enables command input control of either: voltage, current, or dc power. The power supply has analog output as well to assist with process monitoring and for feedback to the controller model in the real-time algorithm. Analog output from all systems presented in Fig. 8.1 was used for feedback to the Simulink Real Time model presented in Fig. 8.2.

The xPC target machine uses an Intel Celeron processor with 32-bit/33 MHz PCI bus. One advantage to HIL testing is that the capability of the processor may be evaluated or even tailored to the requirements of the process under control. For example, the Celeron processor is more than capable of process control computing for the ion-plating process. The Simulink Real Time utilities monitor how hard the processor and PCI bus are working during operation, and this information may be used to specify a lower-cost and less-complex computing system for final machine operation. Another example for tailoring the computing system is for data collection and archival logging of the coating process. The HIL utilities can give a quantifiable requirement for computing capability, bus communications, and memory storage based on algorithm performance on the xPC host.

Figure 8.2 presents the Simulink Real Time process code used for HIL testing of the ion-plating system. Everything to the left of the output module is controlled from within the Simulink Real Time model, which includes PID and LQR controller models for the subsystems. The input module on the right in Fig. 8.2 is used to interface each subsystem of the ion plater with the Simulink Real Time model, including the mass flow controller, plasma power supply, and the conductance valve. For the present study all I/O are analog from 0 to 15 V dc, and no digital communications were used in these HIL tests. The ion flux and peak film stress are calculated as in Eqs. 6.2 and 7.4 using feedback from the manometer and plasma

8.2 Plant Model Characterization

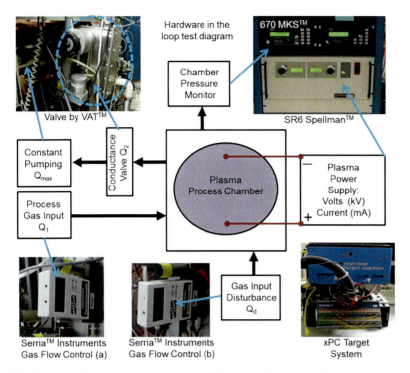

Fig. 8.1 Diagram of the experimental setup for HIL testing of an ion-plating process

power supply. These two calculations were monitored for coating and process health during ion-plater operation.

8.2 Plant Model Characterization

Pressure control inside the ion-plater chamber with input disturbances is the focus of this section. The subsystems of the ion-plating machine from Chaps. 6 and 7 range in age from 2 to 15 years old, representing a huge gap in capability and communications. Yet, this capability gap may be overcome through appropriate control algorithm development. Within a manufacturing setting, it is common practice to merge old and new subsystems of existing deposition equipment. Considering the cost of purchasing new equipment and especially the risk of new equipment installation and qualification, adding new subsystems to an older, yet well-understood, piece of equipment can be more cost-effective.

150　　　　　　　　　　　　　　　　　　　　　　　8　Real-Time Process Control

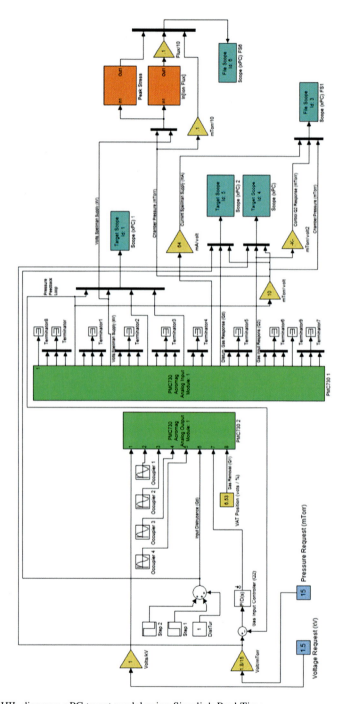

Fig. 8.2 HIL diagram: xPC target model using Simulink Real Time

8.2 Plant Model Characterization

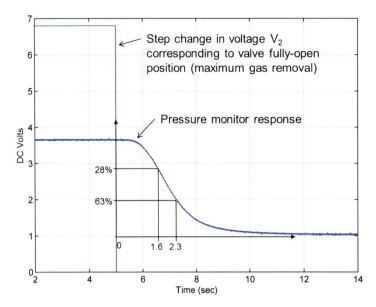

Fig. 8.3 Experimental data to determine the plant model for the chamber. Pressure response to a step-change in valve conductance position starting from 18 mTorr down to 5 mTorr

8.2.1 Conductance Valve Characterization

The first step is to approximate a plant model for each of the subsystems presented in Fig. 8.1 so that a controller scheme may be chosen. A first-order plant model for each subsystem will be used to design a corresponding controller. The subsystem may be understood by measuring system response to a known input, a delta function, or step input, for example. One can apply a step voltage input to the system under investigation and then monitor its output response over time. For example, the response of the conductance valve and chamber pressure related to the exit flow rate Q_2 is presented in Fig. 8.3. The time it takes for the chamber pressure to reach the measurement points, 28 % and 63 %, of the maximum output is recorded on the graph.

A step input to open the valve to maximum position is applied beginning at 5 s. Referring to Fig. 8.3 there is a response delay to the input voltage, and therefore, the first-order plant model must include dead time. The plant model with dead time is structured as

$$\sigma = \frac{e^{-\theta s}}{\tau s + 1}, \qquad (8.1)$$

where the constants θ and τ are calculated as

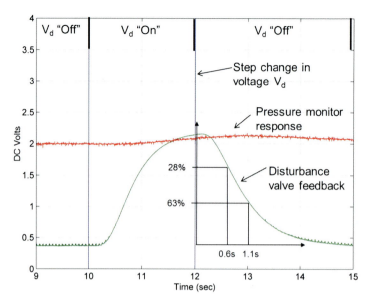

Fig. 8.4 Experimental data to determine disturbance plant model parameters, corresponding to 60–82–60 sccm argon gas flow into the chamber

$$\tau = \frac{3}{2}(t_2 - t_1), \quad (8.2)$$

$$\theta = t_2 - \tau, \quad (8.3)$$

with the values of t_1 and t_2 taken from Fig. 8.3. The plant model in Eqs. 8.1, 8.2, and 8.3 is first order, yet this simple model can give insight when choosing an effective controller algorithm. For example, since the conductance valve has dead time, a fast controller may not be a good choice. Perhaps a regulator controller would be more suited for operation to prevent controller overshoot. Fundamentally, the more time spent with the system measuring first-order parameters, the better understanding of the components you will have. HIL testing allows further adjustment and evaluation of the controller first-order parameters as well.

Measurements for the characterization of the input disturbance plant model Q_d are shown in Fig. 8.4. A step input of 22 sccm argon gas is applied to the chamber system for 2 s then switched off, and the corresponding response of the conductance valve and resulting chamber pressure is measured. Similar plant model information was gathered for the input gas plant model Q_1. All first-order plant models were calculated as described in Eqs. 8.1, 8.2, and 8.3. From the RCF life test results of Chaps. 6 and 7, the process pressure corresponding to the longest RCF life is between 15 and 17.5 mTorr. The inlet and exit flows, Q_1 and Q_2, were set to provide net flow of 60 sccm, which corresponds to approximately 15.5 mTorr steady-state pressure inside the chamber using the cryogenic pump system on the ion plater of Fig. 4.1.

8.2 Plant Model Characterization

Referring to Fig. 8.4, the flow of the step input disturbance test ranged from 60 to 82 to 60 sccm over time. During the test, the flows Q_1 and Q_2 were held constant to maintain 15.5 mTorr pressure and only the disturbance input, Q_d, was increased for 2 s to 22 sccm. There is a 2 s delay in the pressure monitor response as well as an overshoot after the disturbance Q_d is switched off. To summarize, the pressure monitor system was late to respond to the disturbance, which led to a delayed response or the conductance valve to maintain requested chamber pressure. The conductance valve has a delay as well, and the sum of delays is about 4 s, which resulted in chamber pressure overshoot as shown in the pressure monitor response curve in Fig. 8.4.

8.2.2 Manometer and Plasma-Current Response

For comparison, two pressure control schemes were constructed using manometer feedback and plasma-current monitoring. The plasma-current feedback from the power supply was tracked within the HIL code shown in Fig. 8.2 and used as feedback to one of the candidate control schemes. Plasma current is related to chamber pressure as described by Eq. 6.2. Specifically the ion density n_i and the argon gas density n_g influence the plasma-current density. Therefore, a pressure disturbance inside the chamber results in a corresponding disturbance in the plasma current as well. To compare these two controllers, a 2 s, 17 sccm gas input disturbance was applied to the chamber as shown in Figs. 8.5 and 8.6 for each controller type. The controller in Fig. 8.5 uses feedback from the pressure manometer system, the MKS 670 in Fig. 8.1, and the control scheme in Fig. 8.6 uses feedback based on plasma-current measurement from the plasma power supply. Controller action was applied to the inlet flow system, Q_1, while the exit flow Q_2 was held to a fixed set point for all controller tests. Based on characterization of the conductance valve and its response in Fig. 8.3, a constant set-point control of this valve, related to the exit flow Q_2, was most appropriate since its dead-time was on the order of 1.8 s. The plasma-current response presented in Figs. 8.5 and 8.6 is noisy and no filtering was applied so as not to influence the measurement and controller response. However, an isolation amplifier was used to protect the I/O module and host xPC from electrical damage in case of an arcing event during testing.

Observation of the plasma-current response of Figs. 8.5 and 8.6 indicates that plasma-current tracks very similar in shape as the input disturbance, Q_d. From a controls standpoint the desired outcome is that feedback from either the plasma power supply or the manometer would be similar in shape to the disturbance input, which would indicate that these systems are capable of detecting and responding to the disturbance input. Figure 8.5 reveals a delayed response of the pressure manometer to the input disturbance, Q_d. The disturbance is applied between 10 and 12 s, yet the pressure manometer just begins to report the pressure increase

154   8 Real-Time Process Control

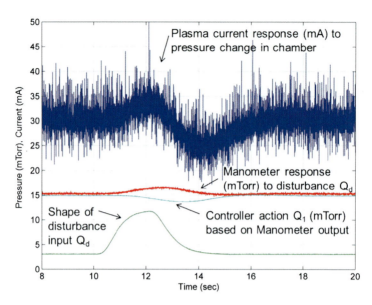

Fig. 8.5 Comparison of manometer and plasma-current response to disturbance input Q_d. Controller action applied to Q_1 using manometer feedback

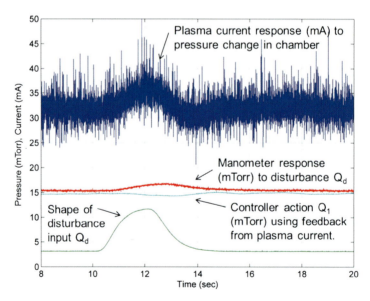

Fig. 8.6 Comparison of manometer and plasma-current response to disturbance input Q_d. Controller action applied to Q_1 using plasma-current feedback

at about 11 s. In comparison, the plasma-current measurement from the plasma power supply tracks the pressure disturbance very closely, almost identical to the shape of the disturbance, Q_d. The plasma current as measured from the plasma power supply is a faster indicator of pressure changes inside the chamber. Alternatively, an in situ Langmuir probe system would provide the same information along with direct calculation of ion density and energy.

The controller of Fig. 8.6 using plasma-current feedback does not cause pressure overshoot. Comparing the plasma-current responses between Figs. 8.5 and 8.6, the pressure reading reported from the manometer caused an overshoot, which would result in a momentary aberration in the coating process. In comparison, feedback from the plasma-current measurement is very fast and there is no overshoot using the control scheme of Fig. 8.6. These results suggest that plasma state variables such as, ion current, ion density, and electron current could all be used for pressure disturbance detection and rejection.

8.3 Process Control Using Plasma Current

Extracting pressure information from a dc-plasma device was first presented in 1965 with the development of ionization vacuum gages (Wheeler and Ganji 2004). An ionization gage is a small plasma chamber that is attached to the chamber of interest. Voltage is applied to a grid that is situated between an anode and cathode configuration within the gage chamber. Ion and electron current are measured, and ion current is the key indicator of gas pressure inside the chamber. The concept of using ion and electron current to measure and track chamber gas pressure may be adapted to a faster plasma measurement tool such as an in situ Langmuir probe system.

The component response and characterization data collected in the previous section is similar for all vacuum and chamber systems related to deposition. For example, ion-plating systems and large coating systems have mass-flow control valves, conductance valves, power supplies, and pumping systems. The disturbance detection and process control challenges are similar for all these types of coating systems. The characterization experiments of Sect. 8.2.1 reveal that manometer-based pressure monitoring is not fast enough to respond to the pressure disturbances presented in this chapter. A typical pressure disturbance may last for 0.5–2 s and change the chamber pressure by as much as 3 mTorr. Direct measurement of plasma current to detect pressure disturbances is a better choice to detect pressure disturbances.

Reference

Wheeler A, Ganji A. Introduction to engineering experimentation. 2nd ed. Upper Saddle River: Pearson Prentice Hall; 2004.

Chapter 9
Closing Chapter: Disturbance Rejection

Abbreviations

M_v	Command input to the physical plant
M_d	Disturbance model feedback to the MPC
M_0	Physical-plant feedback to the MPC
LQR	Linear quadratic regulator
MIMO	Multiple input multiple output
MPC	Model predictive control
N4SID	Numerical Subspace State-Space System Identification
SVD	Singular value decomposition

A survey of the recent literature concerning plasma-based processes reveals that direct feedback of plasma parameters for process control is currently being explored related to plasma-cleaning operations. Feedback from plasma monitoring systems such as in situ Langmuir probes would enable more sophisticated control characteristics such as disturbance rejection and adaptive control. Adaptive control schemes rely on predictive control algorithms that require some knowledge of system behavior. The combination of plasma feedback and adaptive control would enable very good disturbance rejection algorithms.

The goal of this chapter is a discussion of control methods that could be applied to the ion-plating process of Chap. 8 to enhance disturbance rejection. A passive solution for the very slow response of the conductance valve was implemented in Chap. 8, that is, to hold the conductance valve to a fixed set-point control to prevent correction overshoot. An optimization strategy like LQR control could be used as well to improve disturbance response time, as was shown in Figs. 6.12–6.14 of Chap. 6. A more sophisticated control solution would include simultaneous control of the input gas Q_1 with the conductance valve and plasma power supply, based on feedback from ion density and ion current measurements. A multiple input and multiple output (MIMO) controller with predictive capability would give superior disturbance rejection performance.

The predictive nature and the categories of pressure and plasma disturbances that occur during plasma deposition processes would enable predictive knowledge input to a control scheme. Most pressure excursions that occur during deposition are due to molecular dissociation and decomposition of surface contaminants such as light oils, salts, and dust. The response of these contaminants in the plasma sheath is reproducible, which suggests that a predictive control algorithm would mitigate the pressure disturbances caused by the dissociation. Zhang et al. (2011) have shown that the electronic nature of the decomposition of light oils within a dc-plasma may be plotted and studied for controller optimization. In their publication, they describe a process to quantify pressure bursts brought on by contaminants within the decomposition process. A key enabler to more powerful disturbance mitigation would be to trigger controller action based on current disruption within the plasma itself.

9.1 System Identification

Before a predictive controller can be implemented on the coating system, a state-space model estimate of the plant dynamics will be needed. This section is devoted to state-space estimation of physical-plant systems. The purpose of this section is to present the reader with the fundamental aspects of system identification. We include one example calculation from the literature to illustrate the computation process. Unlike the topics in previous chapters, system identification is relatively new within linear control technology. We present the derivations here for clarity to aid understanding of the plant model estimation process that forms the basis of system identification.

System identification is widely used in statistical processing and control of plant-controller systems. The plant is modeled as a linear state-space system similar to

$$\frac{dx(t)}{dt} = Ax(t) + Bu(t), \tag{9.1}$$

with the output response defined as

$$y(t) = Cx(t) + Du(t). \tag{9.2}$$

The matrices $A, B, C,$ and D define the plant model, load, response, and through-pass elements, respectively, of the state-space system. The state variable $x(t)$ corresponds to the operating values of the process control parameters such as voltage, pressure, and plasma current, for example, during model calculation. The output response of the model in Eq. 9.1 is shown in Eq. 9.2 and describes the output of each state with influence from all the other states. For example, we know from previous

9.1 System Identification

chapters that pressure, current, and voltage are related concerning the ion-plating process. The relationship between states is tracked through the plant matrix, A. The influence of the state variables $x(t)$ on the operating parameters is captured in the output response $y(t)$ through the response matrix, C. The input $u(t)$ to the system is similar to the "requested" command input during the coating process. For more information concerning state-space modeling and analysis, we refer the reader to any one of the many textbooks written about this subject such as *Linear Systems* by Decarlo (1989).

System identification may be carried out using the N4SID algorithm implemented in MATLAB. N4SID stands for: Numerical Subspace State-Space System Identification. The N4SID algorithm is a state-space projection method that uses singular value decomposition (SVD), matrix QR factorization, and least squares approximation to estimate the system matrices, A, B, C, and D. The algorithm uses system input and corresponding response output data to construct intra-system matrices that contain the estimated system matrices. For example, the physical system is exercised by applying a known input to all process variables, and then the resulting output response of the physical plant is recorded over time. This process is similar to the plant model characterization method used in Chap. 8 to identify dead time and the plant model of the conductance valve. The main difference with using N4SID is that the input process is carried out at the system level, that is, all plant variables are step input at the same time to track interactions within the physical plant. The N4SID algorithm is more robust and has lower computational complexity compared to other system identification methods. Since it is a subspace projection-based method, it is well suited for multi-input–multi-output (MIMO) systems.

There are several implementations of the N4SID algorithm available in the literature. The N4SID form used in this section is borrowed from Overschee and De Moor (1992) and is summarized here for clarity. The N4SID algorithm may be summarized in five steps:

1. Construct projection matrices and their QR factorizations.
2. Apply SVD to determine estimated system order.
3. Computation estimate of state variables construction.
4. Least squares solution of the intra-system using results from steps 1–3.
5. Extraction of A, B, C, and D matrices from the blocked least squares solution.

Step 1 The input and output data from the exercised system is used to construct the projection matrices. For the derivation presented here, a random input will be used to reduce derivation complexity. Any known input may be used, but we will stick with a random-spike input to keep the derivation simple. The projections are computed as follows:

$$Z_i = \frac{Y_{i,2i-1}}{\left(\frac{U_{0,j-1}}{Y_{0,j-1}}\right)}, \quad Z_{i+1} = \frac{Y_{i+1,2i-1}}{\left(\frac{U_{0,j-1}}{Y_{0,j-1}}\right)}, \quad (9.3)$$

where $U_{0,j-1}$ and $Y_{0,j-1}$ are block Henkel input and output matrices, respectively. The input and output matrices are constructed as

$$U_{0,j-1} = \begin{bmatrix} u_0 & u_1 & u_2 & \cdots & u_{j-1} \\ u_1 & u_2 & u_3 & & \vdots \\ u_2 & u_3 & & & \vdots \\ & & & & \vdots \\ u_{j-1} & u_i & u_{i+1} & \cdots & u_{i+j-2} \end{bmatrix},$$

$$Y_{0,j-1} = \begin{bmatrix} y_0 & y_1 & y_2 & \cdots & y_{j-1} \\ y_1 & y_2 & y_3 & & \vdots \\ y_2 & y_3 & & & \vdots \\ & & & & \vdots \\ y_{j-1} & y_i & y_{i+1} & \cdots & y_{i+j-2} \end{bmatrix}, \quad (9.4)$$

where "i" is the length or size of the input and output vectors. For the derivation, "j" is assumed to be very large, but for actual implementation j is the number of trials carried out using the random input. For example, one could apply a single random input to all states and then record system response. Alternately, multiple inputs and multiple responses could be tracked as well. In addition, a $Y_{i,2i-1}$ matrix could be constructed in the same way as $Y_{0,j-1}$ but using predicted forward output data based on the current-step input/output data. The projections Z_i and Z_{i+1} can also be expressed as $Z_i = \left(L_i^1, L_i^2\right)\left(\frac{U_{0,j-1}}{Y_{0,j-1}}\right)$ and $Z_{i+1} = \left(L_{i+1}^1, L_{i+1}^2\right)\left(\frac{U_{0,j-1}}{Y_{0,j-1}}\right)$ which are used to solve for $L_i^1, L_i^2, L_{i+1}^1,$ and L_{i+1}^2. The matrix divisions of Eq. 9.3 are computed as

$$\frac{U}{Y} = UY^T \left(YY^T\right)^{-1} Y. \quad (9.5)$$

Step 2 Singular value decomposition (SVD) is used to determine the system order. Up to this point, all projections and matrices contain physical-plant information which will be used to determine the system order. The system order is calculated as

$$\left(L_i^1, L_i^2\right)\left(\frac{U_{0,i-1}}{Y_{0,i-1}}\right) = \left(W_{s_1}, W_{s_2}\right)\begin{pmatrix} \sigma_1 & 0 \\ 0 & 0 \end{pmatrix} V^T, \quad (9.6)$$

where the order is equal to the number of nonzero singular values σ_1. The matrices W_s and V are the system column and row space, respectively. Solving for the SVD of the system enables model order reduction as well and is useful to determine

9.1 System Identification

system observability during controller design. The N4SID algorithm will automatically return the reduced order system based on the measured output.

Step 3 The states of the estimated system are computed using the column space and singular values as computed above. The estimated states are computed as

$$X_i = W_{s_1}\sqrt{\sigma_1}\left(\frac{U_{0,i-1}}{Y_{0,i-1}}\right), \quad X_{i+1} = W_{s_{i+1}}\sqrt{\sigma_{i+1}}\left(\frac{U_{0,i-1}}{Y_{0,i-1}}\right), \quad (9.7)$$

where σ_{i+1} is computed using L^1_{i+1} and L^2_{i+1}.

Step 4 The least squares problem to be solved is formulated as

$$\left(\frac{X_{i+1}}{Y_i}\right)^{n+l,j} = \begin{pmatrix} \beta_{11} & \beta_{12} \\ \beta_{21} & \beta_{22} \end{pmatrix}^{n+l,n+m} \left(\frac{X_i}{U_i}\right)^{n+m,j} + \begin{pmatrix} \rho_1 \\ \rho_2 \end{pmatrix}^{n+l,j}, \quad (9.8)$$

where ρ_1 and ρ_2 are related to the error of the fit.

Step 5 Extract the approximated system matrices from Eq. 9.8 as

$$\begin{pmatrix} A & B \\ C & D \end{pmatrix} = \begin{pmatrix} \beta_{11} & \beta_{12} \\ \beta_{21} & \beta_{22} \end{pmatrix}. \quad (9.9)$$

The N4SID summary shown in Eqs. 9.3, 9.4, 9.5, 9.6, 9.7, 9.8, and 9.9 has been implemented into MATLAB and may be used from the command line or from a Simulink model block in combination with a reference controller.

Before using the N4SID algorithm for plant model estimation, it is good practice to test out the estimation process on a known system and to compare results with others. The researchers in Rasmussen et al. (2005) have used N4SID to estimate a plant model of a physical system used in a vapor compression process. In their publication titled *Model-Driven System Identification of Transcritical Vapor Compression Systems*, they use an existing physical plant to generate output for plant estimation study using N4SID. For comparison we will start with the same input and output data presented in their work and compare the results using MATLAB.

Example 9.1: Plant Estimation Comparison from the Literature The goal of this example is to use N4SID to reproduce the results of Rasmussen et al. (2005). Our comparison will be limited to Tables 1 and 2 of their publication. The N4SID command using the MATLAB implementation is given as

$$\text{n4sid}(z, [1:10], \text{'Ts'}, 0), \quad (9.10)$$

where z is a column vector of the input and output data and [1:10] are the model order choices that MATLAB will consider. For example, we can specify model order, or we could ask the algorithm to choose model order from 1 to 10. Using the

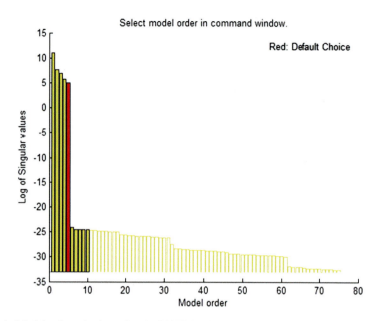

Fig. 9.1 Model order selection using the MATLAB implementation of N4SID from Example 9.1

[1:10] option, the MATLAB implementation will suggest a model order based on the largest number of nonzero singular values in the SVD calculation from step 2 in this section. The parameter "Ts" is the time step for the estimation process, and the value, "0," in the last column of Eq. 9.10 indicates a continuous time model calculation. After the estimation, model order reduction may be selected from the output shown in Fig. 9.1.

A fifth order system is suggested as indicated by the red bar in the graph of Fig. 9.1, which matched the results presented in Table 2 of Rasmussen et al. (2005). Table 9.1 contains a comparison of eigenvalues using N4SID.

The eigenvalues of our plant model estimate and those of Rasmussen et al. (2005) very nearly match based on the reduced order model circled in Table 9.1. As seen in Table 9.1, the plant model estimator will automatically reduce model order based on the SVD calculation and report only those eigenvalues that influence system response. This example demonstrates that the N4SID algorithm implemented in MATLAB is ready for use in the next section for model predictive control.

9.2 Model Predictive Control

Model predictive control (MPC) uses preexisting plant models or estimations of the plant to compute the optimal control law in the presence of known and unknown disturbances. The estimated plant model would be generated using system

9.2 Model Predictive Control

Table 9.1 Comparison of SVD eigenvalues from Rasmussen et al. (2005) and using MATLAB N4SID

Full order model based on physical plant	Reduced order model using system identification	Our model using MATLAB N4SID
−124.020	Eliminated	–
−54.165	Eliminated	–
−28.090	Eliminated	–
−14.598	Eliminated	–
−1.995	−2.202	−2.241
−0.472 ± 0.233i	9.0 −0.518 ± 0.238i	9.1 −0.500 ± 0.237i
−0.175	−0.182	−0.180
0.061	−0.063	−0.063
0.000	Eliminated	–

identification presented in Sect. 9.1. There are two plants within the MPC control scheme: the physical plant under control and the plant model or estimate of the plant model. More plant models may be added concerning disturbances of known physical phenomena. For example, Zhang et al. (2011) have shown that specific types of electrical disturbances can be modeled using linear transfer functions and state-space models. For the ion-plating process of Chaps. 6 and 7, one would construct a library of transfer functions for all types of disturbances that occur during plating. The disturbance models could be theoretical based or could be generated using system identification as well.

The MPC controller receives feedback from the output of the physical plant and the disturbance plant model and optimizes the command input to the physical plant based on the best outcome using the estimated plant model. Using the estimated plant model, the MPC controller compares output from the physical plant with the expected output based on the plant estimation model. If the physical-plant output varies from the expected output by more than the specified tolerance, the MPC controller will adjust the input control law to the physical plant to bring its output closer to the estimation output. A MPC controller model is presented in Fig. 9.2. The requested plant operation, Reference Input, is first given to the MPC controller and it decides what should be sent to the physical plant based on plant output. The inputs M_0 and M_d are the plant output feedback and disturbance model feedback, respectively. The output M_v is the command input to the physical plant.

Concerning the ion-plating process of Chaps. 6 and 7, the requested input is the voltage, pressure, gas flow, and plasma properties sent to the plant to execute a coating process. The primary advantage of the MPC controller is model-reference checking of the physical-plant output with the expected plant output. Another advantage of the MPC controller is that all states are connected through the controller, instead of operating independently of each other as in the control schemes of Chaps. 6 and 7.

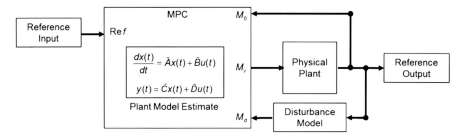

Fig. 9.2 MPC and plant model control scheme with known disturbance model plant function input

9.3 Closing Comments and Future Testing

The MPC controller is well suited for the ion-plating process and the types of disturbances encountered during deposition. All of the disturbances that occur during deposition are identifiable and may be recreated so that one could use system identification to generate a library of disturbance models and then include them into a MPC controller model using Simulink or MATLAB functions. Work is currently underway to improve the disturbance rejection capability of the MPC controller of Fig. 9.2 using the RCF platform of Chap. 5.

References

DeCarlo R. Linear systems: a state variable approach with numerical implementation. Englewood Cliffs: Prentice Hall; 1989.
Rasmussen B, Alleyne A, Musser A. Model-driven system identification of transcritical vapor compression systems. IEEE Trans Control Syst Technol. 2005;13:444–451.
Van Overschee P, De Moor B. N4SID: subspace algorithms for the identification of combined deterministic-stochastic systems. Automatica. 1992;30:75–93.
Zhang Y, Keville BJ, Holohan A, Daniels S. Experimental implementation of robust multivariable real-time feedback control design for RIE plasma processing. AVS 58th international symposium and exhibition. Nashville: Zhang Y, 3 Nov 2011.

Index

A
Abrasive wear, 37, 38
Accelerated testing, 66, 67
Accelerometer, 19–24, 56, 60, 61, 65, 66
Adhesion, 3, 5, 39, 48, 60, 76, 87–95, 100–102, 104, 107, 128, 133, 136, 143, 144
AES. *See* Auger electron spectroscopy (AES)
Analog output, 148
ANSI T5 Steel, 63
Argon Plasma, 1, 89, 92–94, 105, 120
Argon sputtering, 120
ASTM 3359, 93
ASTM-B905, 92
ASTM STP771, 58
Auger electron spectroscopy (AES), 5, 54, 87–89, 92, 95, 135, 136, 138, 139

C
Closed-contact wear, 37
Cohesion, 92, 93, 95
Composite modulus, 40, 42, 43, 45, 57
Constraints, 26–31
Contact radius, 39, 41, 59
Copper gasket, 12, 24, 65

D
Dead time, 118, 151–153, 159
Debye length, 131, 132
Design of experiments (DoE), 107–114, 120, 122, 129–131, 133–142
Disturbance input, 4, 123, 153–154
Disturbance rejection, 2, 5, 157–164
DoE. *See* Design of experiments (DoE)

E
Electron energy distribution function (EEDF), 128, 133, 134
Electrons, 28, 56, 57, 87, 101, 102, 105, 106, 121, 132, 133
Entropy, 50

F
Failure mode, 36, 84, 112–114, 124, 128
Fatigue, 2–4, 14, 17, 18, 33, 36–39, 41, 47–51, 53–84, 92, 101, 103–105, 107, 112, 120
Fatigue wear, 36–38, 48–50
Film
  density, 27
  deposition, 1–3, 100, 102, 118
  stress, 2, 3, 5, 101, 124, 128, 132, 136, 138–144, 148
  thickness, 56, 80, 84, 88, 89, 104, 109
Flexible bellows, 11, 14, 18–19, 25

H
Hardware in the loop testing, 5
Hertz contact stress, 62, 80, 84, 103–105, 108, 112
High voltage, 1–3, 10, 13, 16, 38, 53–55, 94, 101, 122

I
Incipient sliding, 38, 46–48, 78, 79
Interstitial defect, 140
Ion density, 115, 116, 122, 128, 132, 133, 153, 155, 157

Ion Flux, 3, 104, 129, 131–132, 134, 136–144, 148
Ionization, 12, 102, 105, 106, 120, 132, 134, 155
Ionization gauge, 155
Ion mixing, 106, 136, 137, 139, 144
Ion penetration, 138
Ion Plating, 2–5, 32, 46, 90, 99–125, 128, 129, 131, 133, 138, 143, 144, 147–149, 155, 157, 159, 163, 164
I-V curve, 129, 130

L
Langmuir probe, 4, 5, 128–135, 155, 157
Linear quadratic regulator (LQR), 115, 120, 122–125, 148, 157
Lundberg-Palmgren model, 78, 82–83

M
Material hardness, 40
Maxwell-Boltzmann, 133, 134
Mean free path, 115, 131, 132
Mixed-boundary layer lubrication, 56
Model predictive control (MPC), 5, 162–164
M50 Steel, 62, 63, 69, 71, 72, 84, 104
Multiple input multiple output (MIMO), 158, 159

N
Numerical Subspace State-Space System Identification (N4SID), 159, 161–163

O
Objective function, 16, 29
Optimization, 16, 17, 24–30, 122, 125, 147, 157, 158
Oxidation, 13

P
PDE, 140
Physisorption, 56
PID. *See* Proportional-integral-derivative (PID)
Plasma sheath, 3, 4, 101, 105, 106, 110, 111, 114–116, 121, 131, 158
Principal stress, 40

Process
 control, 4, 54, 100, 147–155, 157, 158
 model, 99–125
 parameters, 4, 36, 84, 118, 127–144
Proportional-integral-derivative (PID), 115, 117–120, 122–125, 148

R
Real Time workshop, 147–155
Reliability, 66, 70–72
Residual gas analysis (RGA), 4, 32–33, 56, 112, 141
Rolling contact wear, 46–51

S
Safety interlock, 15–17
Sample calibration, 91
Scanning electron microscope (SEM), 54, 55, 76, 77, 100, 103, 104, 110, 111, 135
Scratch testing, 3, 55, 88
SEM. *See* Scanning electron microscope (SEM)
Sequential quadratic programming, 29
Sheath thickness, 4, 102, 110, 115, 116, 132
Silicon nitride (Si_3N_4), 62–64, 71, 80, 82, 84, 93, 94, 135–136
Silver film, 3, 38, 55, 60, 64, 76, 79, 84
Singular value decomposition (SVD), 159, 160, 162, 163
Solid lubricants, 1, 39, 54–56, 58, 78, 142
Spallation, 58, 112–114, 128
Sputter yield, 111, 116, 118, 122
Stress cycle curve, 68
Surface outgassing, 13, 65
SVD. *See* Singular value decomposition (SVD)

T
Tape testing, 93–95
Thermodynamic stability, 49–51
Thickness by weight, 89–91
Thin film lubrication, 37
Third body transfer model, 77–82, 84
304 Stainless steel, 13, 18, 25, 29, 60, 128
Titanium nitride (TiN), 63, 88, 89, 93, 94, 103, 104

Index 167

U
Ultrahigh vacuum (UHV), 16, 60, 80, 88

V
Vacancy defect, 143
Vacuum chamber, 4, 9–33, 41, 57, 60, 62, 64, 65, 129
Vacuum chamber optimization, 16
Vibration
    frequency, 37, 103
    response, 21, 23, 24, 38, 66
    transmittance, 4, 11, 14, 20, 21

Virtual leak, 14
Viton, 12

W
Weibull distribution, 66, 67, 70
Work hardening, 37, 38, 58

X
X-ray fluorescence spectroscopy, 91, 92, 95
X-ray tube, 1–3, 13, 38, 48, 53–55, 63, 103